D1067546

PROBABILITY IN THE SCIENCES

VOLUME 201

PROBABILITY IN THE SCIENCES

Edited by

EVANDRO AGAZZI

University of Fribourg, Switzerland,
University of Genova, Italy and
President of the International Academy of Philosophy of Science

KLUWER ACADEMIC PUBLISHERS

DORDRECHT / BOSTON / LONDON

Library of Congress Cataloging in Publication Data

Probability in the sciences / edited by Evandro Agazzi.
p. cm. -- (Synthese library ; 201)
Includes index.
ISBN 9027728089
1. Science--Philosophy. 2. Probabilities. I. Agazzi, Evandro.
II. Series.
Q174.P76 1988
501--dc19 88-22993
CIP

ISBN 90-277-2808-9

Published by Kluwer Academic Publishers,
P.O. Box 17, 3300 AA Dordrecht, The Netherlands.

Kluwer Academic Publishers incorporates
the publishing programmes of
D. Reidel, Martinus Nijhoff, Dr W. Junk and MTP Press.

Sold and distributed in the U.S.A. and Canada
by Kluwer Academic Publishers,
101 Philip Drive, Norwell, MA 02061, U.S.A.

In all other countries, sold and distributed
by Kluwer Academic Publishers Group,
P.O. Box 322, 3300 AH Dordrecht, The Netherlands.

Printed in The Netherlands

TABLE OF CONTENTS

PREFACE

Probability has become one of the most characteristic concepts of modern culture, and a 'probabilistic way of thinking' may be said to have penetrated almost every sector of our intellectual life. However it would be difficult to determine an explicit list of 'positive' features, to be proposed as identification marks of this way of thinking. One would rather say that it is characterized by certain 'negative' features, i.e. by certain attitudes which appear to be the negation of well established traditional assumptions, conceptual frameworks, world outlooks and the like. It is because of this opposition to tradition that the probabilistic approach is perceived as expressing a 'modern' intellectual style.

As an example one could mention the widespread diffidence in philosophy with respect to self-contained systems claiming to express apodictic truths, instead of which much weaker pretensions are preferred, that express 'probable' interpretations of reality, of history, of man (the hermeneutic trend). An analogous example is represented by the interest devoted to the study of different patterns of 'argumentation', dealing wiht reasonings which rely not so much on the truth of the premisses and stringent formal logic links, but on a display of contextual conditions (depending on the audience, and on accepted standards, judgements, and values), which render the premisses and the conclusions more 'probable' (the new rhetoric). One could also mention the difficulty with which absolute values and moral standards are accepted nowadays, and the fact that the application of such standards to concrete cases is hardly believed to lead to clear-cut judgements, so that evaluations affected by a certain margin of uncertainty, and therefore implying a 'probabilistic' attitude, are seen as being inevitably needed. In short, we could say that probability appears in such examples as something opposed to the traditional privileged status accorded to 'absoluteness' in different fields.

Examples of another kind express the idea of probability as something opposite to necessity in a rather complex way. Indeed the traditional outlook, developed as a consequence of the growth of modern natural science, had been that of strict determinism dominating the world of nature, paralleled by freedom and indeterminacy characterizing the world of man. But our century has produced something like a reversal of this view: indeterminism has been conceived as one of the fundamental characteristic of nature, while human behaviour has been presented as the result of the operation of a great number of deterministic mechanisms of a genetic, physiological, psychological and sociological nature. The probabilistic way of thinking plays an essential role in both cases: those who advocate a basic determinism need a probabilistic approach in order to account for deviations from the expected regularities, due to insufficient knowledge of the interfering deterministic chains. Those who advocate a basic indeterminism must resort to a probabilistic treatment in order to account for global regularities, as well as statements on single events meaningful and testable. Quantum physics is the most explored - but not the only existing - field where this kind of problem is discussed.

Finally we could mention that probability has also assumed the role of being a replacement for the concept of teleology, or of finalistic order. Indeed, chance has appeared as something which is characterized by certain kinds of regularity, which are different from those expressing 'laws', and from those expressing a design or intended goal. These regularities are mathematically described in the probability calculus and, in such a way, probability has been proposed by many as the 'modern' way of understanding and explaining order and functionality without resorting to teleological principles.

Our list of examples might be easily lengthened, but what has been presented is already enough to show two things. In the first place, if we consider these 'polar oppositions', we are easily lead to ask the question whether they really are as modern as they are claimed to be: indeed they have been clearly present (even though under different forms) in many periods throughout history. Secondly, this characterization of the probabilistic way of thinking through opposition or negation does not show its 'positive' characteristics, which have actually determined its undeniable pre-eminence in modern times. Now these characteristics emerged particularly through a scientific approach to probability, both in the sense of an effort of mathematical definition of this concept, and of a great display of applications in different fields of scientific research. Therefore an exploration of the role of probability in the sciences is important for the intellectual clarification of this concept, no less than for the appreciation of its cultural significance and philosophical relevance. From this point of view, this exploration seems to be even more interesting than the traditional discussions on the so- called foundations of probability, which nowadays appear to have been often conducted with regard to a very abstract and undetermined notion of probability, and without awareness of the fact that the very meaning of probability is deeply affected by the scientific context in which it is applied.

This problem constituted the focus of a conference of the International Academy of Philosophy of Science, which took place at Vico Equense (Naples) in May 1987, in collaboration with the IPE (Istituto per ricerche ed attivitá educative) of Naples. This book contains the invited papers and a selection of the contributed papers presented at this conference, which have been organized into three thematic fields. The first tackles the logical, methodological, and philosophical aspects of probability, both in the sense of being a logical, methodological and philosophical analysis of the concept, and in the sense of examining the role played by probability in the methodology and philosophy of science. The second concerns a few topics where the scientific treatment of probability appears to have given rise to interesting methodological questions, in particular in connection with statistics and information theory. The third sector concerns the application of probabilistic ways of thinking in some of the most classical domains of the natural sciences.

Most of the contributors to this volume are working scientists in different fields, while others are specialists in logic and philosophy of science. This balanced contribution of complementary approaches gives the book a breadth of perspective which may recommend it to a wide spectrum of readers.

Thanks must be expressed here to Dr. Craig Dilworth for his revision of the English style of many papers, and to Dr. Dominique Wisler for setting the camera-ready copy. The publication of this volume has been made possible through a grant of the Italian Consiglio Nazionale delle Ricerche and a financial aid of the Conseil de l'Université de Fribourg.

EVANDRO AGAZZI

PART 1

LOGICAL, METHODOLOGICAL AND PHILOSOPHICAL ASPECTS OF PROBABILITY

Evandro Agazzi

PROBABILITY: A COMPOSITE CONCEPT

1. GENERAL PROPOSALS

The aim of these pages is not to survey the different ways in which probability has been thought of in recent years by the scholars who have contributed to developing one or other of the so-called "theories of probability". Indeed such a survey would be of very little interest, since there already exist excellent books devoted to the presentation and discussion of these theories in their more or less standardized forms (i.e. classical theory, frequentist theory, logical theory, subjectivist theory, propensity theory) and in their variants. Moreover, the discussion of these approaches has shown that each of them has specific merits and specific shortcomings at the same time, so that the debate on the 'foundations of probability' seems to have reached a stage of saturation, comparable with that which occurred in the debate on the foundations of mathematics, where the impression is that the different schools have expressed all their best, without managing to gain the upper hand. The difference might be, that in the case of the foundations of mathematics the new attitude is something like skepticism about the whole foundational enterprise, while in the case of probability the new attitude is rather that of tolerance. Besides, an encouragement toward tolerance has been recognized for a long time in the fact that all probability *theories* lead to the acceptance of the same probability *calculus*, which thus takes on the status of a neutral point of reference, especially after the explicit axiomatization provided by Kolmogorov. As a consequence, the calculus (or the *mathematical* theory of probability) appears as a kind of objective core, with regard to which the other theories

E. Agazzi (ed.), Probability in the Sciences, 3–26.

rather play the role of more or less philosophical interpretations or justifications.

However, this may not be completely true, since there seem to be certain legitimate and rigorous uses of the notion of probability which cannot be adequately mapped onto the rules of the usual probability calculus. Hence an undeniable interest surfaces again for a discussion about the *concept* of probability, yet no longer with the aim, or the pretension, of codifying it in a fixed *definition*, but rather with the aim of *analyzing* it in its different components, such as they result from an inspection of the uses of this concept in various contexts. This is why an interesting clarification may be expected from the study of how the concept (and not just the calculus) of probability is used in the different sciences, or even within broader contexts in which it plays a significant role. It is my view that such an exploration shows the concept of probability to be a quite composite one, containing different semantic components, which have been *separately* privileged by the different 'schools' and promoted by each of them to the rank of *the* definitory characteristic of probability. But since this concept is composite, none of these partial characterizations was able to capture its entire meaning, and this accounts for the impasse we have mentioned above.

My proposal will therefore be that of presenting some elements of the said analysis, by investigating, partly historically, and partly conceptually, some of the most significant contexts in which the notion of probability is used.

2. HISTORICAL CONSIDERATIONS

I have just claimed that considering the *mathematical* theory of probability as being the only solid objective incarnation of this concept might lead to an unjustified restriction of the horizon. This is confirmed by what we find in one of the best books written on probability in the last decades, i.e. in I. Hacking's *The Emergence of Probability*. At the beginning of this book the author says:

Probability has two aspects. It is connected with the degree of belief warranted by evidence, and it is connected with the tendency, displayed by some chance devices,

to produce stable relative frequencies. Neither of these aspects was self-consciously and deliberately apprehended by any substantial body of thinkers before the time of Pascal.[1]

This statement is simply historically wrong, since discussions "connected with the degree of belief warranted by evidence" happened to be widespread in the hellenistic philosophy of the last two centuries B.C. and the first two A.D., i.e. in the philosophy of the 'academic skepticism', whose main representatives were Arcesilaus and Carneades, and of skeptical Pyrrhonists like Sextus Empiricus. The doctrines of these authors had been presented in Cicero's *Academica* and had vigorously surfaced again in the late Renaissance, being discussed, accepted or rejected by scholars such as Montaigne, Descartes, Pascal, Huygens, Bayle, and up to Hume.[2] The fact that Hacking does not take them into account may simply be the result of ignorance, but it has certainly much to do with the fact that the *mathematical* treatment of probability did not really begin before the time of Pascal, so that what had happened before tends to be considered as belonging to the pre-history of the doctrines on probability rather than to its history proper.

Yet this approach contains at least one serious flaw: it makes probability "emerge" with authors who in a way do not give a special importance to this term, and does not appreciate a tradition in which the term "probability" was abundantly and scholarly used, and used exactly in one of the two basic meanings considered as characterizing it (the meaning of a "degree of inference warranted by evidence"). Indeed the titles of the monographs devoted to the mathematical theory of probability did not explicitly mention probability during more than one and a half centuries: think of Huygens *On the ways of reasoning in games of chance* (1657), or of James Bernoulli's *Art of Conjecturing* (1713), or of De Moivres's *Doctrine of Chance* (1759), or of Bayes' *Essay Toward Solving a Problem in the Doctrine of Chances* (1764). It is perhaps only with Laplace that the term is 'officially' adopted as a central notion in the scientific use, with its *Analytical Theory of Probabilities* (1812) and its *Philosophical Essay on Probabilities* (1814).

Hence the question spontaneously arises: what did probability mean in its long history before the creation of a mathematical theory of it? The answer is not difficult, since Cicero uses both the adjective *probabile* and the substantive *probabilitas* and, moreover, he uses them with two different meanings in his different works. In his early rhetorical work *De inventione* the *probabile* is defined as that which "for the most part usually comes to pass, or which is a part of the ordinary belief of mankind"(I,46). Here we find a shade of 'objective' interpretation of probability in the sense of high frequency, but especially a resonance of the Aristotelian concept of *endoxon*, i.e. of that which is "generally believed", or well attested by people. In this sense, "probable" is said of an opinion which enjoys a wide acceptance: this is the meaning preserved in Latin medieval philosophy (e.g. in Aquinas), and Hacking takes this line of tradition into account.

However in Cicero's *Academica* the meaning of *probabile* is different, and translates the Greek *pithanon*, by which the skeptics denoted a subjective attitude based on the evaluation of empirical evidence which provides different degrees of confidence but never certainty, so that the wise man has to live, to use the words of Carneades "using probability as the guide of life". This is so because man is always obliged to be content with "appearances", which always retain the possibility of being deceptive, so that the most he can reach is something "which resembles the truth". In this second approach we see probability in the light of the 'subjectivist' interpretation, i.e. in one of the basic meanings which Hacking mentions without tracing them back to ancient sources.

As a conclusion to these historical remarks we may recognize that the concept of probability seems to be bound to the idea of a certain plausibility of a *state of affairs*, which is translated into the acceptability and verisimilitude of the *judgment* or the *proposition* expressing this state of affairs. Hence the original context of probability is the problem of certainty and truth, which accounts in particular for the etymology of this term in certain languages such as German, where *Wahrscheinlichkeit* etymologically means "verisimilitude". Modern views, such as those of probability as a property of propositions, or of probability as an intermediate value between truth and falsity, or finally of probability as a degree of subjective belief, are already to be found *in nuce* in what we have mentioned.

What might surprise us is that no apparent connection seems to exist in these ancient approaches between probability and forecast, while this connection seems essential to our way of thinking of probability, but there are reasons for this lack.

3. PROBABILITY AND CHANCE

In antiquity, in scholastic philosophy and in the philosophy of the Renaissance before Galilei, as we have seen, probability was a notion concerning statements describing general states of affairs and was conceived as a kind of reasonable guideline for the action of the wise man. This implies two things: that the object of a probability judgment was not a single event, and that the domain of application of probability was not the realm of chance. This does not mean that ancient philosophy did not admit chance in the world: in fact chance - though being rejected as the fundamental mark of reality by the majority of thinkers - was explicitly admitted e.g. by the earliest atomists Leucippus and Democritus, in the sense of an absence of order or design in the universe, while Aristotle admitted the existence of events which are purely "accidental", and the Epicureans even introduced intrinsic randomness in the world by claiming that falling atoms may spontaneously deviate from the vertical trajectory (imposed on them by weight) by a small *clinamen* or "deviation", which at the same time accounts for their giving rise to compound bodies by aggregation, and leaves a space for human freedom. Yet probability, which specifically regarded the acceptability of statements and was expressed by a graduation in the degree of truth or reliability that may be credited to them, was not usually meant to regard random events. Probability and chance were disconnected, the first having to do with the uncertainty which affects our knowledge for various reasons, and the second having to do with the absence of order, regularity or even laws governing the phenomena: in spite of the fact that games of chance have existed in different cultures from times immemorial, we never see them being submitted to probability judgments before Galilei, to whom we can trace back the first documented mention of a problem in a game of chance which is treated by probabilistic reasoning (in the modern mathematical sense[3]). The most interesting confirmation of this distinct status of probability and chance is probably to be found in the works of Pascal (who, incidentally, started the merging of the two domains). In fact in it famous *pari* ("bet") it discusses the problem of the existence of God, which by no means might be considered a matter of *chance*, and he envisages it according to a model derived from the reasonable attitude of a gambler, who has to compare the *risk* of loosing his stake with the size of the possible *win*, and concludes that in the particular question envisaged, the size of the win is incommensurably greater than that of the stake, so that it is absolutely reasonable to accept the existence of God. But this is clearly something which has to do with a *theory of decision* and not with a theory of gambling or of chance, and ranks very naturally along with the traditional line of probability judgments being identical with the most reasonable criteria of assent to statements, upon which the wise man has to regulate his existential choices (as Cicero had clearly explained in his *Academica*). But on another occasion Pascal tackles the problem of how to divide the stakes when a game of chance has to be interrupted, and then he proposes a solution that is really based upon the (mathematical) consideration of chance, since the situation is really a random one.

But this double attitude of Pascal is easily understandable if we take into consideration that in the first half of the 17th century people had become more and more attentive to certain theoretical problems connected with games of chance, and more precisely with the effort of *explaining* certain regularities which appear in the *frequencies* occurring in them, as is easily found in the historical records of these discussions. From this moment on, we can certainly say that a new concept of probability begins to 'emerge', viz. that which is related to the empirically discovered existence of *regularities in chance* (i.e. with the second factor mentioned in the above quoted lines of Hacking's book, "the tendency, displayed by some chance devices,to produce stable relative frequencies").

Some features are worth noting in this connection. First there is a reason for which this kind of problem paved the way to the use of the term "probability" in this context, which explicitly occurred only some time later: the reason is that the mathematical arguments developed in this field appeared to provide ground for *reasonable assent* to be given to certain *judgments* concerning the future outcome of some random events (as we have seen above, this fact is clearly depicted in the titles of some relevant works, where *reasoning* and *conjecturing* are explicitly mentioned). Secondly, probability appears to be more and more connected with *predictions*, and this is on the contrary, rather a new fact. Thirdly, and this is again rather new, probability becomes a property of *single events* and receives therefore a kind of 'objective' flavour, the justification of which is looked for in ontological principles, such as that of "equipossibility", or of "insufficient reason", which were later also labelled as "symmetry" or "indifference" principles. These principles were to a large extent *a priori*, but, and this is a fourth feature, they appeared to be related to the external world by the actual inspection of *relative frequencies* (think of the so-called "empirical law of chance", of the "law of large numbers", of Bernoulli's theorem, and so on) and this shows that some kind of *frequentist* perspective was already present. Finally, and this is the fifth feature, the idea of *expectation* was soon introduced,though being somehow corrected through the qualification of "mathematical expectation", which tended to mitigate the subjectivist meaning of a notion that had been actually evoked as a counterpart to the expectation of a gambler in a game of chance.

This brief survey suffices to show that at the beginning of its mathematical treatment, the concept of probability was indeed a composite one, in which nearly all of its subsequent rather unilateral characterizations are present in a more or less accentuated way, from the objectivist to the subjectivist, from the a prioristic to the frequentist, from the logical to the decisional views. It is therefore not without interest to see why these different aspects came to be split and even opposed to one another, at least in the scientific use of the notion of probability.

4. PROBABILITY AND DETERMINISM

It would be a little ambiguous to say that with the construction of a mathematical theory of probability during the 18th century probability made its entrance into science. Indeed this statement has at least two different meanings: in a first sense it means that probability was treated *scientifically* (i.e. it became the *subject matter* of some science), in a second sense it means that it was used *to make science* (i.e. it became *an instrument* for scientific inquiry). What happened in the 18th century was only the first thing and, as we have seen, amounted to the construction of a *mathematical theory* of probability understood more or less as the typical feature of random events (more precisely, of events at least similar to games of chance). We might even maintain that this corresponded to the generalized enthusiasm of the people of that time for mathematically mastering as many matters as possible, and among these matters, chance was found to be mathematically masterable.

It was only after the end of that century that the newly constructed mathematical theory of probability became an instrument of science, and more precisely of physics, but this was accompanied by a deep conceptual change, after which it was hardly possible to think of probability as "the doctrine of chance", as the earliest authors had depicted it.

In fact the world view underlaying the "new science" of mechanics, so labelled by Galilei and spectacularly developed by Newton and his followers, was a rigorous determinism, according to which the world of all material phenomena is subjected to inflexible laws admitting of no exceptions. These are laws governing the forces acting upon the material bodies *from the outside*, so that no intrinsic hidden properties in them may determine any change in their movements, which might produce unexpected irregularities. Moreover, these laws are of a mathematical nature, and are expressed through differential equations, so that a perfect knowledge of these laws, when accompanied by an exact knowledge of the state of a physical system, should provide the possibility, at least in principle, of absolutely exact predictions and retrodictions concerning the behaviour of this system, by solving the respective equations. This view rejoins the majority of the cosmological conceptions of antiquity and of the Middle ages, according to which chance is excluded from the world, be it because the universe was supposed to be submitted to a strict *necessity* of an ontological character, be it because it was supposed to be governed by God down to the smallest details. The 'modern' scientific outlook had simply replaced the ontological necessity, or the divine providence, by the presence of all pervasive natural laws. But how was it conceivable to apply the "doctrine of chance" to such a world in which chance does not exist?

The answer is provided by a different conception of chance, which is as old as the strict deterministic outlooks we have mentioned: chance is only a word indicating our *ignorance* of the rigidly determined structure of the world and of its events: from the Stoics, to the medieval theologians, to Spinoza, and many others, this is a well known conception, which makes of chance a *subjective* notion. It is with this notion of chance in mind that the mathematical theory of probability may be thought to apply to the study of a deterministic universe. And this actually happened with Laplace. In the famous introductory chapter "On Probability" of his *Philosophical Essay on Probabilities* he states first, as something totally evident and not deserving any proof, the rigorous deterministic conception that we have summarized above and says that a supreme Mind, which were able to know all natural laws, the physical state of all bodies, and which were able to solve all the equations involved, would perfectly predict the motion and the state of whatever being in the world at whatever time. In comparison with this hypothetical supreme Mind, which would be equipped with a *total knowledge*, men lie in an intermediate situation: they are not completely ignorant, but they are far from knowing all the physical laws, the state of all the bodies existing in the universe and their mutual interactions, besides being unable to solve the practically infinite equations involved. Hence, men have to resort to the theory of probability in order to compensate for their *partial ignorance*.

One cannot avoid recognizing that the view of probability implicit in this picture is a *subjectivist* one, so that it is somehow surprising that Laplace presents it as a justification for using in physics a *mathematical* theory of probability that he had himself systematized in a most 'objective' vein in his *Analytical Theory of Probability*. There he starts, in the best tradition of Euclidean rigorism, with a *definition* of probability, which is supposed to be the fountain-head of all the doctrine, and which defines the probability of an event as the ratio of all favorable to all possible cases, provided they are all "equally possible". The development of the theory then follows the familiar path of idealized examples presented as drawings from urns, which is the direct derivation from the original problematic of the games of chance and has remained constant in the textbooks of the so-called "classical" theory or "range" theory of probability at least up to Borel.

This way of proceeding is in a way strange, but it is not really inconsistent: it simply means that, in spite of the fact that nothing happens at random, we can handle those details of reality which escape our exact knowledge *as if* they were at random, since the mathematical theory of probability tells us how to compensate for our ignorance. We may even say more. In the Laplacean conception, the notion of probability still concerns isolated events and, more precisely, their *predictability*. Moreover, the success of the probabilistic predictions is based upon the presupposition of some *order* (viz. the deterministic order), since nothing could be said, in this view, concerning a totally chaotic world: as a matter of fact, the use of probability is a *consequence* of our ignorance, but is *possible* because of a certain degree of *information* we have regarding that order. The idea of probability as a treatment of *randomness* is alien to this approach as far as physics is concerned.

5. FROM THE SINGLE EVENT TO THE COLLECTIONS

The second breakthrough in the application of probability to physics occurred with the creation of the kinetic theory of gases, thermodynamics and statistical mechanics. These events in the history of science are so well known that we feel dispensed from the task of describing them here, and we shall simply make some comments concerning the novelties which they implied.

The conceptual framework remains strictly deterministic, since the billions of molecules which are thought to constitute even a small volume of a gas are supposed to move according to the strictly deterministic laws of Newtonian mechanics - in particular those of impulsive mechanics and of elastic recoil -, so that the ideal Laplacean supreme Mind could in principle know the position and momentum of each molecule and predict its trajectory and momentum at any time. But this knowledge turns out to be not so much impossible, as rather unnecessary or even *irrelevant* to the real aim of the theory, which is by no means that of predicting, even approximately, the behaviour of single molecules. Indeed, the magnitudes that are being considered are something like pressure, volume, temperature, viscosity, specific heat, which can all be measured directly, and are related to each other by certain *empirical laws*. The concept of the chaotic motion of this turbulent crowd is simply a *model* adopted with the view of *explaining* the said empirical laws. The consequence is *not* that we would be better served if we were able to know the behaviour of every molecule (which we unfortunately cannot, owing to our *ignorance*), but rather that we do not need to rely upon such a knowledge, since what matters are the *cumulative* or *collective* effects of all these motions. Therefore, probabilistic reasonings and calculations are introduced here not as remedies for our ignorance, but as tools for mastering *collective* processes, since the empirically testable magnitudes are thought of as being the result of an accumulation of microeffects (e.g. pressure), or as expressing some *average* or *mean* of certain magnitudes such as kinetic energy (e.g. temperature), or the *sum* of them (e.g. heat). And this means that a transition has occurred from the strictly probabilistic treatment of the single events to the specifically *statistical* treatment of *collective events*.

By mentioning statistics we might be inclined to believe that a *frequentist view* was taking supremacy, since statistical methods, implying the use of the mathematical tools of probability calculus, had been known for almost a century in connection with mortality-rates, accidents, and similar collective events in the practice of insurances, which had taken relative frequencies as equivalent to probabilities. But this first impression is misleading, for in the case, let us say, of insurances, the relative frequency was used as an estimation of the probability of the occurence of a *single* event (e.g. as the probability of death of a person of a certain age taking the insurance), while in the case of the kinetic theory of gases or thermodynamics, nobody would ask anything about the magnitudes characterizing a *single* molecule. This is also confirmed by the fact that the *distributions* considered in the case of the insurance practice are *empirically* established ones, taken from tables or records in which individual cases are *counted*, while the distributions used in the said physical theories express *theoretical conjectures* and are proposed (and even adjusted if necessary) in order to *account* for the collective results observed. Nobody has ever thought of *counting* the molecules with the purpose of establishing or even of checking the correctness of these distributions.

Some rather paradoxical consequences may be noted. Determinism at the microlevel is just *presupposed* (essentially for 'metaphysical' reasons or at most in order to preserve the conceptual framework of the Newtonian mechanics which has to serve as the basis for the explanation), but is actually *irrelevant*, while the actual attention is concentrated on the collections, whose magnitudes are connected by deterministic laws. Nothing of the individual microconstituents is really taken into account except their *number*, and they are equipped with a common fictitious property (a *mean*) which none of them really possesses. The connection between the innocent subjacent determinism and the macroscopic apparent determinism is searched for through a seemingly statistical treatment, but at the price of additional strange arrangements: indeed the theoretically introduced distributions are no random, but regular ones, and even fixed ones, since only these are useful for providing the desired explanation.

The impression of a not fully consistent use of the notion of probability that we feel in the presence of these facts is reinforced by the rather bizarre use of the same notion that we find in connection with the second principle of thermodynamics, when it is said, e.g. that it implies that the universe evolves from a *less probable* to a *more probable* state. But the second principle does not mention probability at all, since it simply states that the entropy increases by $\delta Q/t_1$, in the case of two bodies at the temperatures t_1 and t_2 being put in contact, with $t_1 < t_2$. If we look closer at the reasons which suggested the probabilistic interpretation of the second principle, it is possible to see that this has been done essentially in order to discard irreversibility, which is a stranger to Newtonian mechanics, by transforming it into a 'highly probable' but not absolutely valid feature of reality. But by this device a statistical way of thinking, originally based upon the record of frequencies in those cases in which no reference to a law was possible (mortality, accidents, and so on) had been introduced in a domain where deterministic laws were supposed to have full validity.

6. PROBABILITY AND RANDOMNESS

Let us now consider that use of the notion of probability in connection with randomness, which we have seen to be already included in its origins. Randomness indicates a special kind of variability, which is actually observable in *concrete* events and which may take the place of the generic idea of *chance* as a possible subject matter of the theory of probability. In the concept of a random event we first find the classical idea of *contingency* (i.e. the random event is thought of as one which may happen, but might not happen as well), strengthened by an additional requirement: this event may happen as well as not happen even if *the same conditions* are given. This means that not only the ontological concept of contingency, but the more specific concept of non-lawfulness is implied or, if one prefers, that randomness appears as the exact opposite of determinism. This is why, in particular, the very same concepts which exist concerning determinism concern - with the inverted sign - randomness: for some thinkers randomness has no ontological status, and is so reduced to a purely epistemic situation, for others it has such a status, and expresses the fact that reality may be indetermined at least to a certain extent.

We are not interested in discussing this issue here, since in both cases the practical consequence is the same: of a *single* event *nothing* can be said with certainty before it occurs, and this is why the only way left to us is to speak of it in terms of probability or mean. But why is it so? Because some kind of *re-*

gularity emerges if we take into consideration not the single event, but a sufficiently large *collection* of events of its kind. This means that a difference between regularity and lawfulness has to be admitted, a difference which reveals more aspects than one could expect at first sight.

In the first place, a random or stochastic process shows two features: 1. The individual events occur in a way which is *irregular* and hence *unforseable*; 2. Nevertheless some kind of *regularity* emerges in the long run. We could be led in such a way to saying that a random process is characterized by the bizarre feature that individual irregularity combines with a collective regularity. However this is very ambiguous, since the idea of regularity (or irregularity) does not seem to meaningfully apply to individuals. We can say of a shape, of a succession, of a *composition of parts* in general, that it is or looks regular or irregular, but not of an isolated object. On the contrary, of an individual event we can, and do, correctly say that it occurs according to a *law* or not. Hence we should correct our previous statement by saying: random events combine the absence of individually acting lawfulness with a collectively emerging regularity.

That regularity may emerge without lawfulness is therefore clear, i.e. lawfulness is not a necessary condition for regularity. Yet one might believe that lawfulness be a sufficient condition for regularity, but this is not generally true. If we consider the succession of the figures in the calculation of $\sqrt{2}$, or in the decimal expansion of π, no regularity appears, in the sense that we could make no guess about the structure of the sequence by simply *looking* at its already obtained members, while we could predict with total certainty which figure will occur at the n-th place in this succession, for any n, simply because we know the computational *law* according to which the *single* successive figures are generated.

By the above remarks we did not intend to blame the common way of speaking, according to which we often say that events occuring according to a law are *regular*, or show a *regularity*, but wanted to stress that this way of speaking is ambiguous, in that it trivially understands that this regularity is the law itself. The difference results when we proceed to considering collections of events: if these are produced according to a law applied to identical conditions, they are, at least in principle, all identical, so that their succession is singlevalued, rather then regular; from another point of view, as we have seen in the case of real numbers generated by some mathematical law of calculus, the collection of the single steps does not show identical values at every step, but no regularity either.

All this seems to show beyond any reasonable doubt that in the case of randomness we are dealing with a regularity of the collection which *excludes* uniformity and exact predictability, those are the characteristics of lawfulness. Indeed, it is true that the observed regularities in a random process enable us to determine some *mean*, which we use for assigning probabilities to single events of the collection and making some predictions as well. But here we have to be aware that *dispersion* and *deviation* from the mean are equally essential. If *all* events followed the mean, we would immediately say that we are in the presence of determinism, of law, and not of randomness. On the other hand we also require some regularity at the collective level, since only this regularity permits the calculation of means and, by this device, the possibility of deterministically treating *the collection* as a whole.

A last remark may not be out of place. Randomness, we have said, seems to be characterized by an intermediate status between *constancy* or *invariance*, and complete *irregularity*, in that it implies at the same time *variability* and *regularity*. We have also seen that both constancy and irregularity may be the effect of lawfulness. We have a confirmation of this view in a widespread attitude of common sense: when a highly 'improbable' event occurs - i.e. an event which strongly deviates from the mean -, we tend to think that it could not be the effect *of randomness* and look for some *intentional cause* for it. Here we see the tendency toward identifying random with the mean (which is somehow correct), but at the same time excluding deviation from the mean, which is incorrect. However this tacit tendency is not completely absurd, since it corresponds to the fact that total irregularity (and a strong deviation from the mean may appear to be such) is sometimes the indication of an underlying unknown law, as in the case of the decimal expansions of real numbers just mentioned, and these laws have somehow the flavour of 'intentionality' in the sense that they indicate a programme for operating and constructing, rather than picturing the structure of something which is already accomplished, so that the next-coming figure in the expansion cannot be guessed from the inspection of the already existing succession, while it is easily predictable by one who knows the programme or the project of the calculation.

All this may seem rather strange, after all, because it implies that random is somehow a *typical* way of behaving of events, which can even be characterized *a priori*, by the existence of means, of accepted deviations and dispersions, and so on. But this impression of strangeness vanishes if we consider that this characterization was not aprioristic: it was rather the result of the analysis of the distributions of large collections of events, which could in no way be considered as being the effects of some law or of some intention, and which were therefore 'objectively' random at least in a very reasonable sense, such as deaths, accidents, measurement errors, and so on. These events showed some typical distributions (the first to be found was the famous Gaussian distribution) which could *then* be assumed as characteristic of random events. But this way leads to investigate the role of the frequentist view.

7. PROBABILITIES AND FREQUENCIES

We have already noted that a frequentist use of probability was not made in physics even with the kinetic theory of gases, thermodynamics and statistical mechanics, since no *frequencies* in the proper sense were used, but rather artificially conjectured *distributions* based on ingenious hypotheses concerning the physical model of the molecular motions involved. This means that the relationship of probability and frequencies could essentially remain within the framework of the Laplacean and Bernoullian scheme, i.e. within a *deductive* and combinatorial way of reasoning, in which no real 'counting' of the molecules, implying the *empirical* ascertainment of a frequency, was needed or even thinkable. However the limitations of the traditional aprioristic view began to emerge, since the principle of insufficient reason or indifference, with the symmetry properties it implies, were less and less able to cope with physical situations, in which several special conditions had to be taken into account. Yet the real turning point did not surface inside physics, but in biology.

Here consideration of frequencies over large collections of records had became the standard working tool for people accepting the Darwinian approach to the problem of evolution and of the genetic inheritance. This is clear in the case of the founding fathers of modern inductive statistics, such as F. Galton and K. Pearson[4], for whom the notion of probability had became practically synonymous with that of (relative) frequency. But this was after all very natural,since in their field of research the classical tools for evaluating a priori probabilities were of no use: how could one think of determining which are the 'equally possible' cases, the 'favorable cases', the instances of applicability of the indifference principle, and so on (or even something similar to a conceptual physical model) in the case of large animal populations? Here one has to carefully record what is empirically found and try to single out what regularities the frequencies of certain features show.

This new approach also led to a critical reexamination of certain distributions which had been obtained on the basis of theoretical reasonings by the authors of the classical approach and which had been taken by them to hold for actual frequencies as well. This reconsideration led to recognizing that in the empirically observable distributions of random facts, symmetry is rather exceptional, for small asymmetries are very usual and have to be taken into account by indicating new distributions which are typical of them (Pearson found a dozen of such frequency curves). Moreover it was found that even the Gaussian distribution, known under the name of "error function"', did not actually fit the frequencies of the measurement errors with the accuracy it had been credited with. Hence, something very interesting occurred: methods were devised for evaluating the fitness of a theoretically established curve with respect to the actually recorded frequency distribution (the chi-square test is the best known among such tools). The important conceptual significance of this fact is that it introduces into the theory of probability the analogy of the idea of comparing *conceptual experiments* with the outcome of *actually performed experiments*, which was common in physics. But, in spite of that, the use of probability in physics had remained at the level of conceptual experiments, not very different from the conceptual experiments used by classical theorists imagining drawings from urns. It is with the new trend, with the application to large samples, that probability theory acquires the status of an empirical science, or at least of a science having its empirical counterpart in the study of random events, which are no longer characterized by means of a priori established mathematical distribution functions (such as the normal or Gaussian one), but through empirically detected frequency distributions.

8. PROBABILITY AND QUANTUM PHYSICS

It is therefore no wonder that the frequentist approach to probability quickly became predominant, up to the point of inspiring a new *definition* of probability in the well known papers of von Mises.[5] What is of interest in this approach is not so much the 'definition' of probability itself (with its logically rather unsatisfactory concept of a 'limit' of the relative frequencies), as the fact that probability is explicitly and exclusively connected with large collections (von Mises' *Kollektif*) of events, with the requirement of repeatability, and with the additional requirement of randomness of the successions for which the frequencies are calculated. The historical success which this approach enjoyed for a long time was due in part to the great difficulties which the traditional or 'classical' theory was experiencing, not only from a logical point of view (von Mises' criticisms were very effective in this respect), but also from a mathematical point of view (think for instance, of the difficulties connected with the extension of the concept of probability to infinite classes of events, in the well known investigations of Borel). But a major reason for this success was the decisive application that the frequentist interpretation found in quantum physics, where it constituted the ground of the 'orthodox interpretation' of the Copenhagen school.

A difference should be noted with respect to the traditional use of probabilistic methods in classical thermodynamics or statistical mechanics. In that case, as we noted earlier, the real interest was concentrated upon the collections, and collective mean values were used to express macroscopic properties or magnitudes: the behaviour of a single molecule was neither relevant, nor asked about. In the case of quantum mechanics, on the contrary, questions about individual particles make sense again, besides considerations concerning collective events, since an experiment having to do with one single photon is by no means excluded from consideration (it may even give rise to extremely delicate debates, as in the case of the 'two slits experiment'). Now, while in the case of the molecules of classical statistical mechanics it was possible to think that their state be completely determinable in principle, and perhaps possibly determinable in practice through a refinement of the observation instruments, in the case of quantum particles Heisenberg's inequalities showed that such a complete determination was impossible even in principle. The consequence was that the prediction concerning the motion of a single particle was expressible only in terms of probability. Hence we are confronted again with the predictability of individual events, and this prediction is only partially determined, since we cannot avail ourselves of deterministic laws and can only rely upon the estimation of frequencies. This is particularly clear from the interpretation of the meaning of the wave function ψ. This function is regulated by a deterministic equation, but the result of solving it may be interpreted deterministically only in a collective sense (i.e. it gives with full accuracy the frequency of the outcomes of a large number of experiments of the same type), but it may be used as a means for predicting something about a single outcome only in a probabilistic sense, e.g. it enables one to say that a photon with a given frequency distribution is 'potentially' present in a whole finite region, and that the probability of finding it is measured by the frequency distribution itself.

It lies completely outside the scope of these pages to discuss whether quantum indeterminism is an objective or just an epistemic one, and to what extent it may be the consequence of our applying classical concepts to the discription of microobjects. It suffices to note that in the present state of quantum physics, randomness appears as a built-in feature of micro-events and that the frequentist interpretation of probability is the viable way found for treating them adequately.

However we should avoid a very common misunderstanding which is often advocated as a consequence that contemporary physics has become essentially 'probabilistic' or 'statistic'. This statement is sometimes translated by saying that we have finally discovered that physical laws are valid only with a certain probability p. This mistake consists in confusing the probability value p, which a physical law attributes to the occurrence of a given event, and the probability of the law's being valid. Indeed, the law intend to be fully valid, in the sense that it would be falsified if the observed frequency of the event were too different from p, since the law is actually about the collection and not about the single event, so that its occurring or not occurring is *not relevant* to the validity of the law.

9. SMALL SAMPLES

The frequentist view is by no means the last word in matters related to the concept of probability. This approach was based upon the consideration of large collections, which only make a reference to frequency as the basic notion meaningful at all. But what makes the investigations of modern statistics fascinating is that they provide methods for making a reliable use of information derived from small samples. This new trend, in which the most famous name is probably that of Fisher, revealed itself as practically indispensable in the case of experiments performed in laboratories, where high frequencies are out of range for several practical and economical reasons. In these cases a logical study has to be provided as a preparation for the statistical intervention. However it is not the same kind of work as in the case of classical or 'a priori' probability, since the aim is not that of providing a reasonable evaluation of possible and favorable cases, possibly by resorting to ingenious mental experiments, but rather that of inventing a reasonable device for testing some *hypotheses* for which no sufficient empirical evidence, nor cogent ontological arguments are available. This means that in this context a range conception of probability, as well as some subjectivist features are present, but they are connected with an objective frame, which is the *logic of inductive support*. Inductive logic, as was developed by Carnap and other scholars, pertains to this field, but it may be asked whether it was the most fruitful approach. In fact, the very idea of evaluating to what extent empirical evidence supports a statement lead us back to the ancient pre-mathematical idea of probability as a form of reasonable assent to be given to statements on the basis of what experience has taught us. The enthusiastic and astonishingly successful mathematization of this old idea, which occurred in the 18th century in the context of games of chance, has led to the creation of the mathematical calculus of probabilities, but this calculus has proved to be somehow too narrow to cover all the relevant meanings of probability which were implicit in the old idea. Nowadays we have at our disposal certain *logical* calculi of probability which are better suited for treating the problem of inductive support (tink of the well known contributions of Jonathan Cohen[6]), and which do not completely comply with the mathematical calculus of probability. But this should be no wonder: after all, people in the past century found that there are more frequency curves for random events than the classical Gaussian distribution, and in quantum physics there are statistics of particles which cannot be justified on the basis of the classical models of distributing balls in distinct cells. Why should we wonder that there be 'probabilistic' ways of reasoning which are not matched by the conceptualization contained in the probability calculus? The connections between this way of conceiving of probability and decision theory, which are widely recognized at present, are another indication of the legitimacy of this approach.

10. PROBABILITY AND INFORMATION

The last tendencies which we have mentioned, i.e. those in which probability appears again under the form of something connected with a statement rather than with an event or with a collection of events, and having to do with the logic of 'empirically reasonable' arguments, stress a very significant link: that which exists between probability and information. It is exactly this link, which permits the overcoming of a kind of psychological obstacle that made it difficult for rather a long while to attribute a solid scientific status to the probability calculus. In fact, unless one is ready to accept the ontological existence of chance (which was rather difficult to admit in western culture for different reasons we have seen), probability is almost inevitably obliged to be conceived as a tool for coping with our ignorance, i.e., to put it in the words of Cournot, as a "calculus of illusions". It was perhaps owing to reasons of this kind that a scholar as influential as von Mises could write in 1919 that "still to-day the calculus of probability is not a mathematical discipline"[7]. Contrary to these impressions of vagueness one could say that probability calculus provides us with a means for making inferences from what we actually know, which is tantamount to saying that it is conceivable as the study of the best strategies for making the most rational use of our information.

This link is sometimes recognized nowadays in the formal analogies which occur in the Shannonian definition of information measurements and concepts such as those of the 'improbability' of a response, entropy, and so on. Without denying that these analogies might have a euristic value and some intrinsic significance, it seems perhaps most intrinsic to see the realm of probability as something connected with the treatment of 'information' in a less restricted sense than that which is implicit in the technical meaning this term receives within information theory proper.

Another statement one often finds in the literature is that a clear separation between probability "theory" and probability "calculus" has been reached with the axiomatization of this calculus provided by Kolmogorov[8] in the thirties, after which it has become clear that there is a unique probability calculus, which admits of several interpretations or applications. This statement is correct to a certain extent, but it should not be interpreted as expressing the view that the really serious or scientific core of probability is provided by the axiomatized calculus, while the rest are simply philosophical divagations or pragmatic matters. As we have seen, the so-called applications have been historically the very sources from which the meaning of the notion of probability has been derived, and it would not be difficult to see that even some of the most fundamental formal properties of the mathematical calculus of probability were not easily admitted at the beginning, owing to the meaning which was attached to probability (e.g. this was the case with the acceptance of the principle of total probabilities). One does not see any reason that this dynamic development should finish after the creation of axiomatic probability calculus. One should rather be inclined to expect the contrary, since we know that the typical mark of axiomatizations is that of not being absolute and unique. Without going so far as claiming that we should expect to see in the field of probability the creation of fully fledged alternative theories such as those that existed in geometry after the creation of the non Euclidean geometries, we may certainly say that at least some important variants of the traditionally received formal treatment of probability are legitimate, in consideration of the focussing upon aspects of this composite concept which were perhaps not so widely considered or developed in the past, but are becoming more and more so at present.

University of Fribourg (Switzerland)
University of Genova (Italy)

NOTES

[1] See Hacking (1975): 1.

[2] For scholarly backing of these claims I refer to Popkin (1955): 61-71 and (1964). For a discussion of the origins of the probability notion related to this historical account, one may read Jeffrey (1984).

[3] This discussion of Galileo concerned a popular game played with three dices and known as "passadieci". It may be found in the short paper "Sopra le scoperte de i dadi". Published in Galilei (1929-1939), VIII: 591-594.

[4] See Galton (1889), Pearson (1892) and (1948).

[5] See Von Mises (1919).

[6] See Cohen (1970) and (1977).

[7] Von Mises (1919): 52.

[8] See Kolmogorov (1933).

BIBLIOGRAPHY

Cohen, L.J.:

(1970), The Implications of Induction. London: Methuen.

(1977), The Probable and the Provable. Oxford: Clarendon Press.

Galilei, G. (1929-1939), Opere. Firenze: Barbera, Edizione Nazionale.

Galton, F. (1889), Natural Inheritance. London: Macmillan & Co.

Hacking, I. (1975), The Emergence of Probability. Cambridge: Cambridge University Press.

Jeffrey, R. (1984), 'An assessment of the subjectivistic approach to probability', Epistemologia VII, Special Issue: 9-29.

Kolmogorov, A.N. (1933), Grundbegriffe der Wahrscheinlichkeitsrechnung. Berlin: J. Springer. Transl (1956): Foundations of Probability: New York: Chelsea Pub.

Mises, R. von, (1919), 'Grundlagen der Wahrscheinlichkeitsrechnung', Mathematische Zeitschrift, V: 52-99.

Pearson, K.:

(1892), The Grammar of Science. London: Walter Scott.

(1948), Karl Pearson's Early Statistical Papers. Cambridge: Cambridge University Press.

Popkin, R. (1955), 'The skeptical precursors of D. Hume', Philosophy and Phenomenological Research XVI: 61-71.

Popkin, R. (1964), The History of Skepticism from Erasmus to Descartes, New York: Humanities Press (Harper and Row paperbacks, 1968).

Mario Bunge

TWO FACES AND THREE MASKS OF PROBABILITY

The concept of probability has fascinated and puzzled numerous philosophers since its inception three and a half centuries ago. Mathematicians and scientists too, whether basic or applied, have often taken part in philosophical discussions on probability. Notwithstanding such discussions, which have been numerous and often spirited, there is still considerable divergence of opinion concerning the interpretation of probability. (See e.g. du Pasquier 1926, Fine 1973.) This is probably due to the fact that the choice of interpretation is largely a matter of philosophy. Not that philosophy is necessarily inconclusive, but it does colour all thinking on fundamental questions.

Up until one century ago the philosophy of probability was dominated by subjectivism: probability was regarded as a measure of the credibility (or uncertainty, or weakness) of our beliefs. This interpretation had an ontological basis: since the universe was deemed to be strictly deterministic, probability had to be resorted to because of our ignorance of details. (God has no use for probability.) The paradigm cases were the games of chance and the classical kinetic theory of gases: here the basic laws were deterministic but probability was called for because of our ignorance of the initial positions and velocities of the individual entities.

About one century ago an alternative view emerged, namely the frequency interpretation (Venn 1888). According to this view probabilities are long run values of relative frequencies of observed events. While this was a step in the direction of objectivity, it remained half way, because it was concerned with observations rather than with objective facts. Probability was

E. Agazzi (ed.), Probability in the Sciences, 27–49.

regarded as a feature of human experience rather than as a measure of something objective. Like the subjectivistic interpretation, the frequency interpretation is still very much alive - if not *de jure* at least *de facto*.

A third interpretation of probability began to emerge at the time of World War I with reference to statistical mechanics and other stochastic theories, namely the so-called propensity interpretation.According to this view probability values measure the strength of a tendency or disposition of some event to happen. This objectivist interpretation, which had been adumbrated by Poisson and Cournot, can be found in Poincaré (1903), Smoluchowski (1918), Fréchet (1946) and a few others; it was adopted independently by the writer (1951), and has been gaining ground among philosophers, particularly since it was popularized by Popper (1957).

There are then three main views on the nature of (applied) probability: the subjectivist, the frequency, and the propensity interpretations. Up until the birth of quantum theory (1926) the first interpretation was just as compatible with realism as with anti-realism, for one could argue that the basic laws of nature are deterministic, probability being required only because of our empirical ignorance of details. But quantum mechanics and quantum electrodynamics, with their basic stochastic laws, changed the relation of probability to philosophy: from then on the subjectivistic philosophy of probability is compatible only with a subjectivistic philosophy willing to hold that the stochastic laws of quantum mechanics and other scientific theories would cease to hold the day people stopped thinking about atoms, molecules, photons, and other objects with stochastic behaviour.

The frequency interpretation has had a similar fate. While originally it could be espoused by realists as well as by empiricists, ever since the quantum theory was born realists cannot accept it if only because to them the laws of atoms and the like are not supposed to depend upon our observation acts. Thus an atom in an excited state has a definite objective probability of decaying to a lower energy state within the next second, whether or not somebody is counting the actual events of this type in a large assembly of atoms of the same kind. In other words, the propensity interpretation of probability accords well with a realistic interpretation of the quantum theory. But this argument will not persuade someone who is not a realist or who, being a realist, doubts that the quantum theory is here to stay. He will demand more general reasons, i.e. reasons that can be used with reference to all scientific theories.

The purpose of the present paper is to supply such reasons: to show that the subjectivistic and the frequency interpretations are untenable, whereas the propensity interpretation accords well with both the mathematical theory of probability and the stochastic theories of contemporary science. To this end it will prove convenient to start by giving a brief characterization of the theory of whose interpretations are at stake, namely the probability calculus.

1.THE ABSTRACT CONCEPT

Up until half a century ago there was some confusion in the foundations of probability. The confusion consisted in a lack of distinction between the mathematical theory of probability and its various interpretations and applications. So much so that the theory was often presented as if it dealt with physical events. In particular, it was presented as the mathematics of gambling. That stage was overcome by Kolmogoroff's work (1933). This work made it clear that the probability calculus is a branch of pure mathematics -this being why it can be applied in so many different fields of research. Let us give a quick review of the gist of Kolmogoroff's axiom system in its elementary version.

The elementary calculus of probability presupposes ordinary logic (the predicate calculus with identity), elementary set theory, ring theory (a branch of abstract algebra), real analysis, and measure theory. But the foundations of the elementary theory can be understood without the help of any sophisticated mathematics. Indeed the theory has just two basic (or primitive or defining) concepts with a simple mathematical structure. These are the notions of an event (understood in a technical mathematical sense) and of probability measure, which occur in statements of the form "The probability of event x equals y".In principle any set qualifies as an "event"; and the probability of such an "event" is the real number assigned to it by a probability function.

More precisely, a probability function *Pr* is defined on a family *F* of sets such that the union and the intersection of any two members of *F* be in *F*, and also that *F* be closed under complementation. In sum, *F* must be a σ algebra, in the sense that its members obey the laws of the algebra of classes extended to countable infinite unions. This algebric structure is not arbitrary but is demanded by the applications of the calculus. Thus given the probabilities of the complex events x and y, we must be able to compute the probabilities of the complex events "x and y", "x or y" and "not x", and even the probability of an infinite disjunction of events.

Note that in the applications we have to do with *events* proper, not just with abstract sets. But note also that, since real events cannot be negative or disjunctive, the calculus of probability applies to possibilities, not actualities. As soon as any of the events referred to by the expression 'x or y' is actualized, the expression 'the probability of 'x or y' becomes pointless. More precisely, the transition from potentiality to actuality is represented by the transition from $0 < p < 1$ to $p = 0$ or to $p = 1$. See Bunge (1976).

We are now ready for a formal definition of the probability concept, namely thus. Let *F* be a σ algebra on a non-empty set *S* (e.g., $F = 2^S$, i.e. the power set of S) and Pr: $F \rightarrow [0,1]$ a real-valued bounded function on F. Then Pr is a *probability measure of F* if and only if it satisfies the following conditions:

(i) for any countable (finite or infinite) collection of pairwise disjoint sets in *F*, the probability of their union equals the sum of their individual probabilities. That is,if x and y are in *F*, and $x \cap y = \emptyset$, then $Pr(x \cup y) = Pr(x) + Pr(y)$;

(ii) $Pr(S) = 1$.

Note that the theory based on these sole assumptions is *semiabstract* insofar as it does not specify the nature of the elements of the basic set *S* nor, a fortiori, those of the probability space *F*. On the other hand the range of *Pr* is fully interpreted: it is not an abstract set but the unit interval of the real line. Hence the *semi*. Were it not for the semantic indeterminacy of the domain *F* of the probability function, the calculus could not be applied everywhere, from physics and astronomy and chemistry to biology and psychology to sociology. As long as the probability space F is not specified, i.e. as long as no probabilistic model is constructed, probability has nothing to do with possibility, propensity, randomness, or uncertainty.

An *application* of any abstract or *semi*abstract theory to some domain of reality consists in enriching the theory with two different items: (*a*) a model or sketch of the object or domain of facts to which the theory is to be applied, and (*b*) an interpretation of the basic concepts of the theory in terms of the objects to which it is to be applied. Shorter: A factual scientific construct f is a mathematical construct m together with an interpretation I that assigns m a collection P of (really) possible facts; i.e. $f = \langle m, P, I \rangle$. (For details see Bunge 1974.)

In particular, an *application of probability theory* consists in joining the above definition of probability measure (or some of its consequences) with (*a*) a *stochastic model* -e.g. a coin flipping model or an urn model or what have you, and (*b*) a set of *interpretation* (or correspondence or semantic) *assumptions* sketching the specific meanings to be attached to a point x in the probability space F, as well as to its measure $Pr(x)$. As long as these additional assumptions are not introduced, the probability theory is indistinguishable from measure theory, which is a chapter of pure mathematics: only those specifics turn the semiabstract theory into an application of probability theory or part of it.

In other words, the general and semiabstract concept Pr of probability measure is defined (via a set of axioms) in pure mathematics. Each factual interpretation I_i of the domain F of Pr, as well as of the values $Pr(x)$ of the probability measure (for x in F), yields a *factual probability concept* $f_i = \langle\langle F, Pr\rangle, P, I_i\rangle$, where i is a numeral. These various factual probability concepts belong to factual science, not to pure mathematics: they are the probabilities of atomic collisions, of nuclear fissions, of genetic mutations, of survival up to a certain age, of learning a certain item on the *n*th presentation, of moving from one social group to another, and so on and so forth.

What the various specific (or interpreted) probability concepts have in common is clear, namely the mathematical probability theory $<F,Pr>$. This explains why the attempts of the subjectivists and of the empiricicists to *define* the *general* concept of probability either in psychological terms (degrees of belief) or in empirical terms (frequencies of observations) were bound to fail: maximal generality requires deinterpretation, i.e., abstraction or semiabstraction. (For the notions of interpretation and of numerical degree of abstraction see Bunge 1974.)

We can now approach the problem of weighing the claims of the four main doctrines on the nature of (applied) probability.

2. LOGICAL CONCEPT

The so-called logical concept of probability was introduced by Keynes (1921), Jeffreys (1975), Carnap (1950), and a few others. It boils down to the thesis that probability is a property of propositions or a relation between propositions, in particular between hypotheses and the empirical evidence relevant to them.

This view is held almost exclusively by workers in inductive logic, or the theory of confirmation. The center-piece of this theory is a particular interpretation of Bayes's theorem

$$\Pr(h \mid e) = \frac{\Pr(e \mid h) \cdot \Pr(h)}{\Pr(e)}$$

This formula is usually read thus: The probability of hypothesis h given datum e equals the product of the probability of e given h, multiplied by the prior probability of the hypothesis and divided by the prior probability of the evidence.

There are many problems with this reading of Bayes's mathematical correct formula. One is that the only propositions that are assigned definite values are the tautologies and the contradictions, i.e. $\Pr(t) = 1$, and $\Pr(\neg t) = 0$. The remaining propositions, particularly h and e, are assigned probabilities in an arbitrary fashion. For instance, nobody knows how to go about assigning a probability to scientific laws or to scientific data.

The entire enterprise seems to originate in a confusion between probability and degree of truth. The expression "The probability that h be true" is nonsensical (du Pasquier 1926 p. 197). On the other hand proposition "The degree of truth of h equals v", where v is some real number comprised between 0 and 1, does make perfect sense in factual science. In any event, the logical theory of probability has found no applications in science or in technology. In these fields one assigns probabilities to states and events -e.g. to an excited state of an atom, and to the probability that the atom decays from that state to a lower energy state.

Since the "logical" interpretation has been absorbed by the more popular subjectivistic (or personalist) interpretation, we shall turn to the latter.

3. PROBABILITY AS CREDIBILITY

The subjectivistic (personalist, Bayesian) interpretation of probability construes every probability value $Pr(x)$ as a measure of the strength of someone's belief in the fact x, or as the accuracy of his information about x. (Finetti (1972), Jeffreys (1975), Savage (1954).) There are a number of objections to this view.

The first objection, of a logical nature, was raised towards the end of Section 1, namely that one does not succeed in constructing a *general* concept by restricting oneself to a specific interpretation. However, a personalist might concede this point, grant that the general concept of probability belongs in pure mathematics, and claim just that the subjectivist *interpretation* is the only applicable, or useful, or clear one. However, this strategy won't save him, for he still has to face the following objections.

The second objection, of a mathematical nature, is that the expression '$Pr(x) = y$' makes no room for a subject u and the circumstances v under which u estimates his degree of belief in x, under v, as y. In other words, the elementary statements of probability theory are of the form '$Pr(x) = y$', not '$Pr(x, u, v) = y$'. And such additional variables are of course necessary to account for the fact that different subjects assign different credibilities to one and the same item, as well as for the fact that one and the same subject changes his beliefs not just in the light of fresh information but also as a result of sheer changes in mood. In sum, the subjectivist or personalist interpretation of probability is *adventitious*, i.e. incompatible with the mathematical structure of the probability concept.

Even if the former objection is waived aside as a mere technicality -which it is not- a third objection is in order, namely this. It has never been proved in the psychological laboratory that our beliefs are so rational that in fact they satisfy all of the axioms and theorems of probability theory. Rather on the contrary, there is experimental evidence pointing against this thesis. For example, most of us experience no difficulty in holding pairs of beliefs that, on closer inspection, turn out to be mutually incompatible. Of course the subjectivist could go around this objection by claiming that the "calculus of beliefs" is a normative theory not a descriptive one. He may indeed hold that the theory *defines* "rational belief", so that anyone whose behaviour does not conform to the theory departs from rationality instead of refuting the theory. In short he may wish to claim that the theory of probability is a theory of rationality -a philosophical theory rather than a psychological one. This move will save the theory from refutation but it will also deprive it of confirmation.

A fourth objection is as follows. A belief may be construed either as a state of mind (or a brain state) or as a proposition (or statement). If the former then the probability $Pr(x)$ of belief x can be interpreted as a measure of the objective strength of the propensity or tendency for x to occur in the given person's mind (or brain). But this would of course be just an instance of the objectivist interpretation and would be totally alien to the problem of the likelihood of x or even the strength of a subject's belief in x. On the alternative construal of beliefs as statements -which is the usual Bayesian strategy- we are faced with the problem of formulating objective *rules* for assigning them probabilities. So far as I know there are no such (unconventional) rules for allotting probabilities to propositions. In particular, nobody seems to have been able to assign probabilities to scientific hypotheses -except of course arbitrarily. Surely the subjectivist is not worried by this objection: his whole point is that prior probabilities must be guesstimated by the subject, there being no objective tests, whether conceptual or empirical, to estimate the accuracy of his estimates. But this is just a roundabout way of saying that personalist probability is just a flight of fancy that must not be judged by the objective standards of science.

Our fifth objection is but an answer to the claim that probability values must always be assigned on purely subjective "grounds", i.e. on no grounds whatever. If probability assignments were necessarily arbitrary then it would be impossible to account for the scientific practices of (*a*) setting up stochastic models of systems and processes and (*b*) checking the corresponding probability assignments with the help of observation, measurement, or theory. For example, genetic theory assigns definite objective probabilities to certain genetic mutations and recombinations, and experimental biology is in a position to test those theoretical values by contrasting them with observed frequencies. (On the other hand nobody knows how to estimate the probability of either data or hypotheses. We do not even know what it *means* to say that such and such a statement has been assigned this or that probability.) In sum, the subjectivist interpretation of probability is at odds with the method of science: in science (*a*) states of things and changes of state, not propositions, are assigned probabilities, and (*b*) these assignments, far from being subjective, are controlled by observation, measurement or experiment, rather than being arbitrary.

Our sixth and last objection is also perhaps the most obvious of all: if probabilities are credences, how come that all the probabilities we meet in science and technology are probabilities of *states* of concrete things -atoms, molecules, fields, organisms, populations, societies, or what not- or probabilities of *events* occurring in things of that kind, no matter what credences the personalist probabilist may assign either the facts or the theories about such facts? Moreover, many of the events in question, such as atomic collisions and radiative transitions, are improbable or rare, yet we cannot afford to dismiss them as being hardly credible.

The personalist might wish to rejoin that, as a matter of fact, we often do use probability as a measure of certainty or credibility, for example when we have precious little information and when we apply the Bayes-Laplace theorem to the hypothesis/data relation. However, both cases are easily accounted for within the objectivist interpretation, as will be shown presently.

Case 1: incomplete information concerning equiprobable events. Suppose you have two keys, A and B, the first for your house and the second for your office. The probability that A will open the house door is 1, and the probability that it will open the office door is 0; similarly for key B. These are objective probabilities: they are physical properties of the four key-lock couples in question. Suppose now that you are fumbling in the dark with the keys and that you have no tactual clues as to which is which. In this case the two keys are (empirically) equivalent before trying them. Whichever key you try, the probability of *your choosing* the right key for opening either door is 1/2. This is again an objective property, but not one of the four key-lock pairs: it is an objective property of the four you-key-lock triples. Of course these probabilities are not the same as the previous ones: we have now taken a new domain of definition of the probability function. And surely the new probability values might be different for a different person, e.g. one capable of distinguishing the keys (always or with some probability) by some tactual clues. This relativity to the key user does not render probability subjective, any more than the relativity of motion to a reference frame renders motion subjective. Moreover, even when we assign equal probabilities to all the events of a class, for want of precise information about them, we are supposed to check this hypothesis and change it if it proves empirically false. In short, incomplete information is no excuse for subjectivism.

Case 2: inference with the help of the Bayes-Laplace theorem. This is of course the stronghold of the personalist school. However, it is easily stormed. First, recall that the Bayes- Laplace theorem is derivable from the mere definition of conditional probability without assuming any interpretation, whether personalist or objectivist. (Indeed, the definition is: $Pr(x \mid y) = Pr(x \cap y) / Pr(y)$. Exchanging x and y, dividing the two formulas, and rearranging, we get the theorem: $Pr(y \mid x) = Pr(x \mid y) \cdot Pr(y) / Pr(x)$.) Secondly, since there are no rules for assigning probabilities to propositions (recall our fourth objection), it is wrong to set x = evidence statement (e), and y = hypothesis (h) in the above formula, and consequently to use it as a principle of (probabilistic or statistical) inference. However, if we insist on setting $x = e$ (evidence) and $y = h$ (hypothesis), then we must adopt an *indirect* not a literal interpretation: $Pr(h)$ is not the credibility of *hypothesis h* but the probability that the *facts referred to by h* occur just a predicted by h. $Pr(e)$ is the probability of the observable *events described by e*; $Pr(h|e)$ is the probability of the *facts described by h, given* -i.e. it being actually the case- *that the events referred to by e occur*; and $Pr(e|h)$ is the probability of the *event described by e, given that the facts referred to by h happen*. This is the only legitimate interpretation of the Bayes-Laplace theorem because, as emphasized before, scientific theory and scientific experiment allow us to determine only the probabilities of (certain) facts, never the probabilities of propositions concerning facts. A byproduct of this analysis is that all the systems of inductive logic that use the Bayes-Laplace theorem interpreted in terms of hypotheses and data are wrong-headed.

The upshot of our analysis is that the personalist interpretation of probability is wrong.

4. PROBABILITY AS FREQUENCY

If we cannot use the subjectivist interpretation then we must adopt an objectivist one. Now, many objectivists believe that the only viable alternative to the personalist interpretation is the frequency interpretation. The latter boils down to asserting that '$Pr(x) = y$' *means* that the relative long run frequency of event x equals number y or, rather, some rational number close to y. (See Venn (1888), Mises (1972), Wald (1950).)

A first objection that can be raised against the frequency interpretation of probability -and a fortiori against the identification of the two- is that they are *different functions* altogether. Indeed, whereas Pr is defined on a probability space F (as we saw in Section 1), a frequency function f is defined, for every sampling procedure R, on the power set 2^{F^*} of a finite subset F^* of F -namely the set of actually observed events. I.e.

$Pr : F \rightarrow [0,1]$ but $f: 2^{F^*} \times \Pi \rightarrow Q$, where $F^* \subseteq F$, and where Π is the set of sampling procedures (each characterized by a sample size and other statistical parameters) and Q is the set of proper fractions in [0,1].

Our second objection follows from the former: a probability statement does not *refer* to the same things as the corresponding frequency statement. Indeed, whereas a probability statement concerns usually an individual (though possibly complex) fact, the corresponding frequency statement is about a *set* of facts (a "collective", in Mises' terminology) and moreover as chosen in agreement with certain sampling procedures. (Indeed, it follows from our previous analysis of the frequency function that its values are $f(x,\pi)$, where x is a member of the family of sets 2^{F^*} and π a member of Π.) For example, one speaks of the frequency with which one's telephone is observed (e.g. heard) to ring per unit time interval, thus referring to an entire set of events rather than to a single event -which is on the other hand the typical case of probability statements. Of course probabilities can only be computed or measured for event types (or categories of event), never for unique unrepetable events such as my writing this page. But this does not prove that, when writing '$Pr(x) = y$',we are actually referring to a set x of events: though not unique, x is supposed to be an individual event. In other words, whereas *probability statements speak about individual events, frequency statements speak about strings of observed (or at least observable) events*. And, since they do not say the same, they cannot be regarded as identical.

To put the same objection in a slightly different way: the frequency interpretation of probability consists in mistaking percentages for probabilities. Indeed, from the fact that probabilities can *sometimes* be *estimated* by observing relative frequencies, the empiricist probabilist concludes that probabilities are identical with relative frequencies -which is like mistaking sneezes for colds and, in general, indicators for unobservable facts. Worse: frequencies alone do not warrant inferences to probabilities: by itself a percentage is not an unambiguous indicator of randomness. Only a selection mechanism, whether natural or artificial, *if random*, authorizes the interpretation of a frequency as a measure of a probability. For example, if you are given the percentage of events of a kind, and are asked to choose blindfold any of them, then you can assign a probability to your *correctly choosing* the item of interest out of a certain reference class. In short, the inference goes like this: *Percentage & Random choice* ---→ *Probability*. The line is broken to suggest that this is not a rigorous, i.e. deductive inference, but just a plausible one.)

Surely not all frequencies are observed: sometimes they can be calculated, namely on the basis of definite stochastic models, such as the coin flipping model or Bernoulli sequence. (As a matter of fact all the frequencies occurring in the calculus of probability are theoretical not empirical.) But in this case too the expected frequency differs from the corresponding probability. So much so that the difference is precisely the concern of the laws of large numbers of the probability theory. One such theorem states that, in a sequence of Bernoulli trials, such as coin flippings, the frequency f_n of successes or hits in the first n trials approaches the corresponding probability (which is constant, i.e. independent of the size n of the sample). Another theorem states that the probability that f_n deviates from the corresponding probability p by more than a preassigned number tends to zero as n goes to infinity.(Note that there are two probabilities and one frequency at stake in this theorem.) Obliterate the difference between probability and frequency, and the heart of the probability calculus vanishes.This then is our third objection to the frequency interpretation of probability, namely that *it cannot cope with the laws of large numbers*. For further technical objections see Ville (1939).

Our fourth argument is of an ontological nature, namely this.While a frequency is the frequency of the *actual* occurrence of facts of a certain kind, a probability may (though it need not) measure the *possibility* of a fact, or rather the strength of such a possibility. For example, as long as the coin we flipped is in the air, it has a probability 1/2 of landing "head". But once it has landed potentiality has become actuality. On the other hand, the coin has no frequency while it is in the air, for a frequency is a property of the collection ("collective") of the final states of the coin. Consequently identifying probabilities with frequencies (either by definition or by interpretation) implies (*a*) rejecting real or physical possibility, thus forsaking an understanding of all the scientific theories which, like quantum mechanics and population genetics, take real possibility seriously, and (*b*) confusing a theoretical (mathematical) concept with an empirical one.

The correct thing to do with regard to the probability- frequency pair is not to identify them either by way of definition or by way of interpretation, but to clarify their mutual relation as well as their relations to the categories of possibility and actuality. We submit that frequency *estimates* probability, which in turn measures or *quantitates* possibility of a kind, namely chance propensity (Bunge 1976). And, while probability concerns possibles, frequency concerns actuals and moreover, in the applications, it always concerns observed actuals.

In other words: there is no valid frequency *interpretation*, let alone *definition* of probability; what we do have are statistical *estimates* of theoretical probability values. Moreover frequencies are not the sole estimators or indicators of probability. For instance, in atomic and molecular physics transition probabilities are often checked by measuring spectral line intensities or else scattering cross sections. And in statistical mechanics probabilities are estimated by calculating entropy values on the basis of either theoretical considerations (with the help of formulas such as Boltzmann's) or measurements of temperature and other thermodynamic properties. In short, probabilities are not frequencies and they are not interpretable as frequencies although they can sometimes (by no means always) be estimated with the help of frequencies.

To be sure frequencies, when joined to plausible random mechanisms, supply a rough indication of probability values and serve to check probability values and serve to check probability calculations. Hence probabilities and frequencies, far from being unrelated, are in some sort of correspondence. In fact (*a*) an event may be possible and may even have been assigned a non-vanishing probability without ever having been observed to happen, hence without being assigned a frequency; (*b*) conversely, certain events can be observed to occur with a certain frequency without however being assigned a non vanishing probability.

In sum, the frequency interpretation of probability is inadmissible for a number of technical and philosophical reasons.Let us therefore look for an interpretation of probability free from the fatal flaws of the frequency interpretation.

5.PROBABILITY AS PROPENSITY

A mathematical interpretation of the probability calculus, i.e. one remaining within the context of mathematics, consists in specifying the mathematical nature of the members of the domain F of Pr, e.g. as sets of points on a plane, or as sets of integers, or in any other way compatible with the algebraic structure of F.Such an interpretation of the probability space F would yield a *full mathematical interpretation* of the probability theory. (Likewise, interpreting the elements of a group as translations, or as rotations, yields a full mathematical interpretation of the abstract theory of groups.) Obviously, such a mathematical interpretation is insufficient for the applications of probability theory to science or technology. Here we need a factual interpretation of the calculus.

A *factual interpretation* of probability theory is obtained by assigning both F and every value $Pr(x)$ of Pr, for x in F, factual meanings. One such possible interpretation consists in taking the basic set S, out of which F is manufactured, to be the state (or phase) space of a thing. In this way every element of the probability space F is a bunch of states, and $Pr(x)$ becomes the *strength of the propensity or tendency the thing has to dwell in the state or states* x. Similarly, if x and y are states (or sets of states) of a thing, the conditional probability of y given x, i.e. $Pr(y|x)$, is interpreted as the strength of the propensity or tendency for the thing to go from state(s) x to state(s) y. This then is the *propensity interpretation of probability*.

This is not an arbitrary interpretation of the calculus of probability. Given the structure of the probability function and the interpretation of its domain F as a set of facts (or events or states of affairs), the propensity interpretation is the only possible interpretation in factual terms. Indeed, if F represents a set of facts, then $Pr(x)$, where x is in F, cannot but be a property of the individual fact(s) x. That is, contrary to the frequency view (Section 3), probability is *not a collective* or ensemble property, i.e. a property of the entire set F, but a property of every *individual* member of F, namely its propensity to happen. What *are* ensemble properties are, of course, the normalization condition $Pr(S) = 1$ (recall Section 1) and such derived (defined) functions as the moments (in particular the average) of a probability distribution, its standard deviation if it has one, and so on. (This consideration suffices to ruin the frequency school, according to which probability is a collective or ensemble property.)

This point is of both philosophical and scientific interest.Thus some biologists hold that, because the survival probability can be measured only on entire populations, it must be a global property of a population rather than a property of each and every member of the population. (Curiously enough they do not extend this interpretation to the mutation probability.) The truth is of course that, while each probability *function Pr is* a property of the ensemble F, its *values* $Pr(x)$ are properties of the members of F.

It is instructive to contrast the propensity to the frequency interpretations of probability values, assuming that the two agree on the nature of the probability space F. (This assumption is a pretence: not only frequentists like von Mises but also Popper, a philosophical champion of the propensity interpretation, have stated that facts have no probabilities unless they occur in experimentally controlled situations. In fact they emphasize that probabilities are mutual properties of a thing and a measurement set-up -which of course makes it impossible to apply stochastic theories to astrophysics.) This contrast is displayed in the following Table.

$p=Pr(x)$	Propensity	Frequency
0	x has (almost) nil propensity	x is (almost) never observed
$0<p<<1$	x has a weak propensity	x is rarely observed
$0<<p<1$	x has a fair propensity	x is fairly commonly observed
$p\approx 1$	x has a strong propensity	x is very commonly observed
$p=1$	x has an overpowering propensity	x is (almost) always observed

Note the following points. First, although a probability value is meaningful -i.e. it makes sense to speak of an individual fact's propensity- it is so only in relation to a definite probability space F (e.g. with reference to a precise category of trials). Likewise a frequency value makes sense only in relation to a definite sample- population-sampling method triple. For example, the formula "x is rare" presupposes a certain set of occurrences, to which x belongs, among which x happens to be infrequent.

Secondly, in the case of continuous distributions, zero probability is consistent with very rare (isolated) happenings. That is, even if $Pr(x) = 0$, x may happen, though rarely as compared with other events represented in the probability space. Consequently a fact with probability 1 can fail to happen. (Recall that any set of rational numbers has zero Lebesgue measure. Entire sets of states and events are assigned zero probability in statistical mechanics for this very reason even though the system of interest is bound to pass through them. This is what 'almost never' is taken to mean in that context, namely that the states or events in question are attained only denumerably many times.)

Thirdly, the frequency column should be retained alongside the propensity interpretation though in a capacity other than interpretation or definition. Indeed, although the frequency column fails to tell us what "$Pr(x) = y$" *means*, it does tell us under what conditions such a formula is *true*. Long run frequency is in short a *truth condition* for probability statements. Besides, frequency statements have a heuristic value. For example, if p means a transition probability, then the greater p, the more frequent or common the transition.

Fourthly, note again that the present propensity interpretation differs from Popper's in that the latter requires the referent to be coupled to an experimental device. No such hang-up from the frequency (or empiricist) interpretation remains in our own version of the propensity interpretation. Nor do we require that only events proper (i.e. changes of state) be assigned probabilities, as an empiricist must -since states may be unobservable. States too may be assigned probabilities, and in fact they are assigned in many a stochastic theory, such as statistical mechanics and quantum theories. (The statistical mechanical measure of entropy is a function of the thermodynamic probability of a state -or, as Planck put it, it measures the preference [*Vorliebe*] for certain states over others.) In other words not only transition probabilities (which are conditional) but also absolute probabilities can be factually meaningful.

Fifthly, note that the propensity (or any other) *interpretation* of probability is to be distinguished from the probability *elucidation* (or exactification) of the intuitive or presystematic notion of propensity, tendency, or ability. In the former case one attaches factual items to a concept, whereas in the latter one endows a factual concept with a precise mathematical structure. In science (and also in ontology) we need both factual interpretation and mathematical elucidation.

Sixthly, the propensity interpretation presupposes that possibilities can be real or physical rather than being just synonymous with our ignorance of actuality. On the other hand according to the frequency interpretation there is no such thing as a chance propensity for an individual fact (state or event): there would be only limiting frequencies defined for entire ensembles of facts, such as a sequence of throws of a coin, or a family of radiative transitions of a kind.Indeed, the phrase '$Pr(x) = y$' is, according to the frequency school, short for something like 'The relative frequency of x in a large ensemble (or a long sequence) of similar trials is observed to approach y'. This view is refuted by the existence of microphysical theories concerning a single thing, such as a single atom, to be sharply distinguished from a theory about an aggregate of coexisting atoms of the same kind. Another example: in principle, genetics is in a position to calculate the probability of any gene combination - which, given the staggering number of possibilities, is likely to be a one- time event. A relative frequency is a frequency of actuals, hence it cannot be identical with a possibility (measured by a probability).Unlike frequencies, probabilities do measure real (physical) possibilities of a certain kind. Therefore if we take real possibility seriously, i.e. if we are possibilists rather than actualists, we must favor the propensity over the frequency interpretation.

Seventhly, just as the subjectivist interpretation of probability is a necessary constituent of the subjectivist interpretation of quantum physics, so a realistic interpretation of the latter calls for an objective interpretation of probability. For example, the formula for the probability of a radiative transition of an atom or molecule, induced by an external perturbation, is always read in objective terms, not in terms of credences. So much so, that it is tested by measuring the intensity of the spectral line resulting from that transition.Another example is the superposition principle. If a quantum- mechanical entity is in a superposition

$$\Psi = c_1 \cdot \Phi_1 + c_2 \cdot \Phi_2$$

of the eigenfunctions Φ_1 and Φ_2 of an operator representing a certain dynamical variable (e.g. the spin), then the coefficients c_1 and c_2 are the contributions of these eigenfunctions to the state Ψ, and

$$p_1 = |c_1|^2 , \; p_2 = |c_2|^2$$

are the objective probabilities of the collapse of Ψ onto Φ_1 or Φ_2 respectively. This collapse or projection may occur either naturally or as a result of a measurement, which in turn is an objective interaction between the entity in question with a macrophysical thing (the apparatus). Unless the objective interpretation of the above probabilities is adopted, a number of paradoxes may result. One of them is the so-called Zeno's quantum paradox. According to it, if an unstable quantum system is observed continuously to be, say, in state Φ_1 (excited or not-decayed), then Ψ cannot raise its head: it will continue to lie on Φ_1. But of course this result (an interpretation of von Neumann's measurement postulate) runs counter experience, so there must be something wrong with the reasoning. What is wrong is, among other things, the adoption of the subjective interpretation of states, hence of probabilities, as being nothing but states of the observer's knowledge. The paradox dissolves upon adopting the propensity interpretation and taking the superposition principle seriously (Bunge and Kalnay 1983).

In short, there are a number of reasons for favoring the propensity interpretation of probability over its rivals.

6. CONCLUDING REMARKS

We have examined five concepts of probability, each of them in three respects: mathematical validity, scientific viability, and philosophical plausibility. The concepts in question are the following: the semiabstract concept defined implicitly by probability theory, the "logical" concept, the notion of personalist (or subjective or Bayesian) probability, the frequency conception, and the propensity interpretation.

We have taken the majority view that the mathematical concept of probability is adequately characterized by the standard theory of probability developed along the lines of Kolmogoroff's work. The logical conception cannot be implemented and is never used in science.Both the personalist and the frequency notions turn out to be mathematically untenable because at variance with the standard theory of probability. The personalist concept is mathematically invalid because the probability measure defined in the calculus of probability makes no room for any persons. And the frequency view is mathematically incorrect because (*a*) the axioms of the probability calculus do not contain the (semiempirical) notion of frequency, and (*b*) some of the key theorems of the theory of probability, such as the laws of large numbers, concern the differences between probability values and frequencies. Only the propensity interpretation of probability was found mathematically unobjectionable.

As for the use of the various probability concepts in science and technology, the semiabstract concept cannot be applied without further ado: it must be turned into a concept with a factual meaning. This is done by interpreting the basic space S (out of which the probability space F is constructed) in terms of factual items such as the states or the changes of state of a concrete thing. Consequently an arbitrary value $Pr(x)$ of a probability function, for x in F, means the weight or strength of the state(s) x, or else the tendency or propensity for event(s) x to happen. This is the meaning to be assigned to statements of the form "$Pr(x) = y$" occurring in the stochastic theories of pure and applied science. Such statements are objective and in general also testable.

On the other hand science has no use for the personalist view precisely because it is subjective. Who but a dogmatist or a biographer could be interested in pronouncements such as "Bayesian *X* attaches credence *Y* to theory *Z*"? As for the frequency view, it is not viable in science either, because it conflates calculated probabilities with observed frequencies, thus preventing the latter from discharging the function of testing the former. However, the frequency conception has at least some heuristic power, which the personalist does not. For example, if p is the viability (survival probability) of organisms of a certain kind, present in number N, then pN may be interpreted as the fraction of surviving organisms -although strictly speaking pN is only the average number of survivors. We may be permitted to reason that way provided it helps and no traces of such heuristic props remain in the end. And they must not remain if only because actuality (such as the actual fraction of survivors) should not be confused with possibility (as measured by the most probable number of survivors). Besides, the propensity concept is at least as heuristically fertile as the frequency misconception.

The propensity interpretation of probability is then the only one that fits the mathematical theory of probability and is also adaptable to science. Moreover, it is the only one that fits in with a realistic theory of knowledge, whereas the Bayesian view is consistent only with a subjectivist epistemology, and the frequency view invites an empiricist theory of knowledge. Finally, both the Bayesian and the frequency views presuppose classical determinism of the Laplacean style: they equate possibility with conceptual possibility and they deny the reality of chance or randomness. On the other hand the propensity view takes real possibility seriously and admits the reality of chance or randomness. If a concrete thing -be it atom, organism, or community- has propensity $Pr(x)$ to be in state(s) x, or to experience change(s) x, then this is a property the thing possesses independently of our beliefs- a property that can sometimes be checked by observing actual frequencies but must not be confused with the latter.

In short, the propensity interpretation of probability is consistent with the standard theory of probability and with scientific practice, as well as with a realist epistemology and a possibilist ontology. Hence it solves the old tension between rationality and the reality of chance. None of its rivals has these virtues.

McGill University (Canada)

BIBLIOGRAPHY

Bunge, Mario:

(1951), 'What is chance?', Science & Society 15:209-231.

(1974), Interpretation and Truth. Dordrecht & Boston: D. Reidel Publ. Co.

(1976), 'Possibility and probability'. In W.L. Harper & C.A. Hooker, Eds., Foundations of Probability Theory, Statistical Inference, and Statistical Theories of Science, Vol. III, pp. 17-33. Dordrecht-Boston: Reidel.

(1977), The Furniture of the World. In Treatise on Basic Philosophy, vol. III. Dordrecht & Boston: D. Reidel Publ. Co.

(1981), 'Four concepts of probability', Applied Mathematical Modelling 5:306-312.

Bunge, Mario and Andrés J. Kálnay (1983), 'Solution to two paradoxes in the quantum theory of unstable systems'. Nuovo Cimento 77B: 1-9.

Carnap, Rudolf (1950), Logical Foundations of Probability. London: Routledge & Kegan Paul.

du Pasquier, G. (1926), Le calcul des probabilitités, son évolution mathématique et philosophique. Paris: Hermann.

Fine, T.L. (1973), Theories of Probability: An Examination of Foundations. New York: Academic Press.

Finetti, Bruno (1972), Probability, Induction and Statistics. New York: John Wiley.

Fréchet, Maurice (1946), 'Les définitions courantes de la probabilité'. Reprinted in Les mathématiques et le concret, pp. 157-204. Paris: Presses Universitaires de France.

Jeffreys, Harold (1975), Scientific Inference, 3rd ed. Cambridge: Cambridge University Press.

Keynes, John Maynard (1921), Treatise on Probability. London: Macmillan.

Kolmogoroff, Alexander (1933), Grundbegriffe der Wahrscheinlichkeitsrechnung. Berlin: Springer-Verlag.

Mises, Richard von (1972), Wahrscheinlichkeit, Statistik und Wahrheit, 4th ed. Wien: Springer-Verlag.

Poincaré, H. (1903), La science et l'hypothèse. Paris: Flammarion.

Popper, Karl R. (1957), 'The propensity interpretation of the calculus of probability and the quantum theory'. In S. Körner, Ed., Observation and Interpretation. London: Butterworths Sci. Publ.

Reichenbach, Hans (1949), Theory of Probability. Berkeley & Los Angeles: University of California Press.

Savage, Leonard James (1954), The Foundations of Statistics. New York: Wiley.

Smoluchowski, Marian von (1918), 'Über den Begriff des Zufalls und den Ursprung der Wahrscheinlichkeitsgesetze in der Physik'. Naturwissenschaften 6:253.

Venn, John (1888), The Logic of Chance, 3rd ed. London: Macmillan & Co.

Ville, Jean (1939), Etude critique de la notion de collectif. Paris: Gauthier-Villars.

Wald, Abraham (1950), Statistical Decision Functions. New York: Wiley.

Italo Scardovi

AMBIGUOUS USES OF PROBABILITY

I

An urn containing a hundred thousand little balls is set in front of me. I can perceive no difference between them and I am told that each bears a number of the natural series from one to a hundred thousand. I draw out a ball and read the number: whatever it is, the result leaves me indifferent. I notice that the ball is black and I am told that all the balls in the urn were white except for one: the one I have in my hand. I cannot help but be surprised at a result which all schools of probability, as well as common sense, would judge as highly improbable, attributing just one chance to the extraction of a black ball and ninety-nine thousand nine hundred and ninety-nine chances to the opposite occurence. And yet, it is still the same ball, the same event, a single fact: a fact which had initially left me without reaction. Thus, I ask myself "Does a phenomenon, even if casual, assume different meanings only because of a change in the system of reference, the key of interpretation?" On reflection, I realize that also the number extracted itself had only one chance against ninety-nine thousand nine hundred and ninety-nine. I should therefore have been immediately surprised by this too; and also by the extraction of any other number.

What is it, then, that lends significance to an event if not our expectation? Thus, there is a distinction to be drawn between the formal conventions for measuring probability and the suggestive components within the relationship with random events. Such events play within phenomena occurring in nature and life, impressing, at the very roots, their becoming: within the temporal dimension, the snare of chance has become part of all pheno-mena, from elementary particles to the galaxies,

E. Agazzi (ed.), Probability in the Sciences, 51–66.

from nucleotids to man. It offers an intuition of reality which is now essential to a knowledge founded on statistical laws: the laws of a science which is no longer deterministic and transforms probability into a universal heuristic instrument.

Even if we imagine the world as an urn to be interpreted by science, the approaches change according to the theoretical system implied. So much so, that even a simple lotto urn defies all efforts to find an agreement. In fact, the game of lotto seems to provide the starting point for an amusing "little dialogue on minimal systems" (systems for winning), among the three characters taking part in Galileo's famous and delightful "Dialogo": all three bent on predicting the future, on questioning the urn which holds it mistery and poses the conditions.

One of them (Salviati perhaps) maintains that the urn has no memory and therefore each number has the same probability of being extracted every week, thus implying a sort of principle of indifference. Another (Sagredo perhaps) is ready to counter this within the historical fact of delayed numbers and the logico-empirical model of the law of large numbers, thus asserting the validity of betting on delayed numbers. The third (Simplicio, of course) mistrusts all theoretical systems and is unable to accept anything which is different from what he has seen in the past (numbers extracted more times than others). His only line of argument is as follows: "I do not know why this number has been extracted more often than others and more often than is foreseen by your rules: it must have its own good reasons. I do not know what they are, and yet I appreciate a certain facility: therefore, I won't bet on the delayed number, but on the winning one". We may ask ourselves "Is this choice, which seems to contain an animistic component, really the most unfounded, the most absurd? Is this not perhaps what we are doing when we estimate probability on the basis of frequency or when we project a statistical datum beyond the bounds of cognition?"

The answer lies in the empirical and logical significance of "large numbers", i.e. in the tendential and somewhat inertial stability of "statistical laws": the laws of a natural philosophy of no-longer totalitarian propositions, laws of scientific knowledge which has observed casuality to be at the very roots of phenomenal variability. This is the vision of nature which accompanies biological evolutionism, thermodynamics, classic and po-pulation genetics and quantum mechanics: a context in which all sciences have learned to consider "populations" and identify their statistical properties, observing events free from any form of individual necessity as collective phenomena in which immanent accidentality plays a role.

II

If today's biology regards evolution in terms of changes in the allelic frequencies of a population, natural selection as the differential reproductivity of genotypes, genetic drift as chance innovation, this is because biological science has by now become irreducibly statistical and its phenomena are plunged into fortuitous variability: a variability which is the origin of life and which is constantly renewed by life.

The sources of this variability -from spontaneous mutation to the exchange of chromatid segments, from meiotic division to the recombination of crosses- are firmly embedded within the accidental; in its stabilizing and convergent action in large numbers and in its divergent and erratic effect on small numbers. The inner processes of those phenomena -molecular, atomic, subatomic- conceal the ineffable mystery of a nature which really seems to be engaged in a dice game, the Rosetta stone holding the secrets of both life and man. It is emblematic that the casual factors of genetic variability have been baptized by several perplexed immunologists as "Generator of diversity", giving the acronym: God. Anatole France wrote: "Chance, definitely, is God".

In Monod, chance is a factor which disturbs a repetitive mechanism, an accidental occurrence which, by upsetting the order, saves the species; while, in Dobzhansky, the accidental cooperates with the necessary. Nevertheless, evolution remains a non-linear, non-finalistic path: a process, according to Jacob, which operates like a "bricoleur" who assembles fragments intended for different purposes. It is the indeterminate nature of the "bricolage" which makes the immanent accidental and which, according to Prigogine, means that life is no longer an event arising from an alien universe but a "statistic miracle" within the system, an evolutionary effect of the thermodynamic process in a history woven by the necessity of laws and the casuality of fluctuations: fluctuations violating the average tendencies at branching off points.

Anticipating a scientific revolution, Maxwell wrote that the true logic "for" this world is the probability calculus; Boltzmann initiated a methodology which made probability the logic "of" this world, daring to draw his "scandalous" parallels between gambling and the physics of molecular aggregates and concluding that "the probabilistic hypotheses reflect a state of nature". Today's physicist is guided by a statistical intuition of reality, with roots in the quantic interpretation of Copenhagen, dictated by Bohr, in the indetermination principle formulated by Heisenberg (a relation between mean values) and in Born's interpretation of Schroedinger's wave function in a probability distribution.

The probability distribution is essential to the quantic microworld: a statistical reality, where matter is transformed into field, and field intensity -the squared module of the wave function- measures a probability: the probability of finding, within a given space, the observable constituent of "the wave of probability", i.e. the quantic particle. Even more objective is the immanent indeterministic spontaneity of the elementary processes of a life which is objective and irreducibile: a life which is also irreducible for Spinoza's "Deus sive natura", for Laplace's "intelligence assez vaste", for Maxwell's astute little devil, for the microscope of molecular biology. But not for the statistician's "macroscope", which distinguishes distributions whose predictive values do not concern single events but, rather, their probability. He provides the intellectual tools of a scientific methodology which has grown up abstracted from individual events, perceiving statistical collective properties.

Science is not a question of guessing the next genotype; it is an attempt to identify the probability distribution (statistical law) of possible genotypes. It is not a bet on which atom will disintegrate but the search for a decay process (statistical law) of an aggregate of radioactive atoms; it does not anticipate a molecular trajectory but deduces a collective result (again, a statistical law) of the movements in a molecular population. It does not predict an evolutionary pattern but identifies the algorithms (and they are statistical algorithms) of the elementary biophysical processes determining the becoming of life. Even if all the coordinates of these phenomena were known in a given moment, it would be impossible to deduce their state at a different instant: only the probability distribution of the possible states, and nothing else, could be deduced.

III

So what of the individual event? How can the fate of a particle or an allele be predicted? All questions without a theoretic answer if not in terms of "statistic determinism", without an operative answer if not in terms of "statistic probability", understood to be the reduction to an individual level of a group property: a real number, from zero to one, aimed at grading the expectation of a single occurence. It is the probability of the experimental sciences, an abstraction from the population to its individual components. It can be seen as the realization of a renunciation, but a conscious choice based on the statistical intuition of nature.

Without this type of intuition, the natural sciences would never have achieved their most important concept, the concept of evolution. A deterministic theory of biological evolution would produce a one-way, necessary progression, with everything implicit in the original data. Individual variability would be devoid of sense and it would make nonsense of virtual variability compatible with the genome of a species. In such a vision, two genetically identical populations in uniform environments would have to evolve in a parallel fashion; whereas, in the statistical vision, they do not remain identical because genetic casuality (the non-invariant duplication of nucleic acids, independent chromosome assortment, the diploidal result in the zygote) will inevitably lead to their differentiation in the combinatorial and casual becoming of the generations. (However, a staunch determinist would maintain that their divergence was the result of different initial conditions. The sceptical indeterminist would retort that, even if they had been the same, an eventual diversity would still emerge. Two irreconcilable philosophies.).

If biological evolution is not predictable, this is not so much due to our intellectual limitation as to the intervention of accidental processes which are the very origin of variability and do not admit arcane predestinations, determinate necessities or immanent purposes. It is genetics that has revealed the indeterminism of the tiny events giving rise to variability, identifying in nucleic acids the algorithms of combinatory alternatives within linear polymers containing a vast number of possible outcomes, only a minimal part of which are realized - again, by chance- by each species. It is biophysics that has recognized the error arising in DNA duplication to be a consequence of the second law of thermodynamics, and the alteration of the chemical composition of molecules to be an elementary event subject to the principle of indetermination. These are the source of the statistical intuition of the phenomenon of life which has made Darwin's evolutionism and Laplace's determinism definitively incompatible.

Thus, we have a phenomenic conception of chance, as being natural, objective and intrinsic: a more "essential" idea of chance than that which causes the meeting between the hammer of Dubois, the tinsmith, and the head of Dupont, the doctor, in Monod's example, also found in Poincaré and Bergson -chance as destiny, the accidental as fatality (an idea going back to Cournot). However, in the coming together of two separate deterministic sequences, there exists the condition of independence, not one of theoretic unpredictability. Thus, given a specific "cause" of a punctiform mutation, it is not impossible to foresee its effects in theoretical terms. None the less, there exists the contingency on an episodic connection between the initial microevent and its macroscopic consequences within the organism.

It is intrinsic casuality which leads a molecular population towards a growing entropy, to ensure the variability of phenotypes with genotypical variety, to place the unpredictability of an individual event within a form of statistical determinism which transcends it. It is the casuality which Mendel noticed in the genealogies of peas, individually unpredictable and capricious, yet easily resolvable with surprisingly regular statisti-cal schemes. Boltzmann perceived it in the second law of thermodynamics, transforming it into a prototype statistical law. Von Bortkievicz found it in the distribution of fatal horse kicks in the Prussian army. The same intrinsic chance, the same immanent lottery governing the tendential division of stochastic fleas on the two long-suffering dogs of Paul and Tatiana Ehrenfest. Not certain results, merely probable. It is the role of large numbers which makes them come true.

Therefore, probability becomes the intellectual tool of a new rationality: a statistical rationality. But in which theoretical framework? Within deterministic phenomenology, probability is the lack of information, the cognitive limits of the human mind, "ignorance of the true cause of events", so defined by an entire body of scientific thought, from Hume to Planck, from Laplace to Einstein. According to indeterministic phenomenology, probability is in nature: in irreversible mixes, thermal agitation, genetic variability, the elementary phenomena of matter, the evolution of living forms. Where Quetelet envisaged a mysterious lottery transposing onto an individual level a transcendent necessity dictated by a superior collective imperative - the commandment of a God who does not disdain gambling- today's scientist sees collective necessity as an empirical consequences of the irriducible accidentality of elementary events.

IV

In discussing the objective, subjective, cognitive or strategic meaning of probability, we have to confront the underlying issue: chance. Is it an easy game of thought or a mysterious game of nature? An epistemological or an ontological category? The basic meaning of probability cannot but change as a function of the answer. In common speech, probability is already an ambiguous and variable term, overloaded with interpretations and connotations: it is used to refer to the degree of belief in a proposition which cannot be asserted or denied with any certainty; it can also refer to the fortuitous event's potential to occur. There is a profound antinomy between the two meanings: one thing is the subjective choice between a true or false alternative without univocal criteria, another thing is the statistical measure of the immanence of an uncertain event. The confusion of terms, ideas and schools of thought arises from precisely such a failure to distinguish between deterministic and indeterministic assumptions, between subjective uncertainty and objective indeterminacy, the refusal to distinguish (not even in the sense of Carnap) between the meanings attributed to probability and, even worse, from inappropriate distinctions. Fortunately, science is relatively unaffected by nominalistic controversies: it is concerned with the probability of phenomena. In scientific practice, statistic frequencies identified in inertial systems are interpreted as inductive probabilities, Carnap's "probabilities of science". (All elementary probabilities of classic and population genetics are statistical frequencies from which other probabilities are drawn by applying the sum and product principles formulated by Laplace).

To the subjectivist, however, the statistical identification of probability seems too restrictive. He sees probability in terms of belief, a personal expectation of the single event: an expectation determined in terms of betting, thus taking the mathematics of the uncertain back to its very origins. Having been derived from games of chance, probability calculus thus returns to a mere gamble: a bet then, a bet now. (In fact, we already find Kant returning to this concept in the second volume of "Kritik der reinen Vernunft" (1781).

The subjectivist definition of probability as an expression of uncertainty is certainly all-comprehensive. However, we may ask ourselves what place the bet criterion has in scientific thought which, from the late 17th century to the beginning of the present one, underwent one of its most radical revolutions, i.e. the discovery of chance in the elementary phenomena of life and the basic constituents of matter, the establishment of an indeterministic intuition of reality and the recognition of statistical laws alongside, or substituting laws regarding single events. Whereas subjectivism identifies probability as a wholly personal evaluation, scientific realism set out to base probability on phenomena. Thus, this probability is a statistical properties rather than a betting stakes. After the decline of classic determinism, the scientist is left asking Poincaré's question: "Is chance really any different from the name we give to our ignorance?"

To the subjectivist, scientific procedure appears a "superstitions" basis for a logic of uncertainty. And yet, these superstitious probabilities have been the foundation for a new concept of biology and physics. The laws of these sciences are statistical laws and the probabilities inherent in such laws are the result of a process of abstraction, without which no knowledge can exist. Subjectivists view each event as a "unicum" and emphasize the conventional character of "objectivistic" criteria which base probability on an empirical sample. It is true that the categorical repetitivity of the world's facts is a classificatory abstraction, linguistic artefact, semantic allegory. But it is also true that rational knowledge began when facts where reduced into classes (as symbols of conceptual categories), with the search for laws which transcend the unique event. Whereas Leonard Savage asserts that probability must refer to the single case in which a decision is involved and, as such, cannot be resolved within the frequentist definition, the physicist and genetist can reply that the single case, in itself, does not make science: it is, rather, a particular case of an operative choice, of a utilitarian gamble.

Bertrand Russell points out that, all knowledge we possess is either "knowledge of particular facts" or "scientific knowledge". In fact, neither Galileo's artefact, typified by the experiment, nor the statistical event, referring to populations, can be called "particular facts". By considering each event to be unique, probabilistic subjectivism runs the risk of falling into a form of exaggerated solipsistic idealism and becoming isolated from science: a science which has matured by abstracting "the singular", dealing with variability, identifying statistical group properties, with probability distribution the arrival point of experimental process. Herein lies the statistical foundation of scientific probability; and, perhaps, this is where we find the solution to the divergence between a concept of probability based in a phenomenology arising from natural variability and the one depending on the activities connected to the man's hedonistic choices; between a context of knowledge and that of decision. Although a heated debate, it seems to by-pass the important role of the inductive and hypothetical-deductive principles of statistics in the growth of knowledge; it seems to ignore the contribution of the statistical "modus operandi" in the progression of scientific thought, which tends more and more to make the statistical intuition of reality its own "modus intellegendi"; thus, a natural philosophy eroding the rational universe set up by scientific thought on the basis of Newton's "Principia" (1687) - a world system whose apology was provided by Pierre Simon de Laplace, little suspecting he was in fact dictating its epitaph.

V

The knowledge of the innate casuality of the entire microworld has freed scientific thought from a deterministic methodology aimed at codifying the single event, its causal necessity and the predictability of its recurrence. The existence of individual variability offers new connotations to the statistical inductive process, at the same time giving a meaning to statistical probability. Whether it is considered a projection into the unknown of an empirically acquired constant or transformed -by individualization- into a Bernoullian probability or an "a posteriori" probability according to Bayes-Laplace, this process of induction has its primary origin in the large number equilibrium, in the semantic arrow between frequency and probability. It is precisely the inverse probability theorem to prove the supremacy of experimental data over "a priori" assumptions where they are not prejudicial extremes. A methodological model of inductive inference, the rule of dialectic confrontation between idea and fact. And it is the asymptotic irrelevance of initial probability -demonstrated by von Mises- which outlines the much-discussed role of general knowledge, "demonstrates" the inductive supremacy of statistic law and offers new connotations of Hume's problem -a problem which loses its traditional rigidity when inference is derived from statistical properties, probabilistic propositions and tendency laws.

From Bernoulli to Bayes, Leibniz to Laplace, the induction of the degree of the potential occurrence of an event whose frequency is known through a series of observations has taken on a coherent formal character. In Thomas Bayes' "Essay toward solving a problem in the doctrine of chances" (1763, posthumus), the author reduces to a subordinate probability calculus the processes of acquiring experience, merging observations and opinions, and of modifying them according to the facts that occur. In "Mémoire sur la probabilité des causes par les événements" (1774), Laplace formulates this inductive principle in causal terms: after the occurrence of event E, necessarily dependent on one among a number of incompatible circumstances, the probability that one (given) among them has acted is proportional to the product of the probability of the said circumstance multiplied by the probability that, if the circumstance arises, the event is produced. This gives rise to the essential algorithm of inductive inference, the probabilistic scheme defining the semantic link between experience and reason: the formulation of a "final" probability from an initial one with a factor of likelihood.

The Bayes-Laplace formula reflects the role of "a priori" probability, showing its adaptability to the touchstone of experience. The history of scientific methodology is one with the long debate on the role of preliminary knowledge and pre- existing paradigms (whether theoretical systems, working hypotheses or preconceptions...): a radical rethinking of the dialectic relationship between reality and paradigm, empirical world and conceptual scheme, fact and fantasy, and a continual illusion regarding the possibility of an abstract codification of the euristic relationship between theoretical scheme and empirical evidence. However, the relevance of general knowledge cannot help but be subject to change in keeping with historical transformations.

Rather than offering actual rules, or a recipe to replace other recipes, the formula of inverse probability serves to demonstrate that facts win out over preliminary positions -the emergence of the statistical law, in large numbers- in cases where other alternatives are not automatically excluded: when, that is, the "a priori" unitary probability is not all attributed to a single hypothesis. Indeed, the mind of the species appears to be modelled on a large numbers. This is an attitude produced by the experience of phylogenesis and ontogenesis, also liable to Hume's sceptical question which would be posed today in new terms:

"What rational foundation has the empirical regularity of large numbers?"

There is only one answer: chance. It is difficult to say how much that is not rational is concealed by this enigmatic word, defined by Cassirer as a chamaleon, always taking on new colours. There are those who (like Planck, Schroedinger, de Broglie and Einstein) consider chance to be a provisional convention while waiting to decipher "hidden variables" and others who (like Bohr, Heinsenberg, Born, Pauli) view it as an intellectual "point of no return". But the "man of the abacus", like myself, looks to the most ancient and vivid definition: "Nothing which is due to chance occurs for something". Aristotle said as much twenty-four centuries ago. However, he was far from any idea of an empirical pattern in the casual event, of an unsuspected natural order in the results of purposeless occurrences. Such an order was to make itself manifest to the author of "theorem H", when he observed the entropic time pattern of reversible and unpredictable molecular mechanics, within a superior thermodynamic regularity which was both predictable and irreversible.

VI

Thus, a new natural philosophy emerges: the philosophy of a science which deals with populations of particles, molecules, cells, living organisms, in an effort to identify their statistical properties. The methodological framework of this science is still the one outlined by Galileo, what can be called "the hypothetical-deductive canon". However, by providing new semantics, variability and chance impose a new syntax. The incumbent hypothesis is the so-called "null hypothesis" (Ho) i.e. the hypothesis of the totally accidental nature of an observed result (E). This is the first hypothesis to be confronted by the researcher since experimental investigation has started to deal with variable phenomena. But, the probability p(E/Ho) that a result may occur by chance is radically different from the probability p(Ho/E) that a given result is attributable to chance.

A simple formal relation put forward in the Bayes-Laplace scheme:

$$p(Ho/E) = \frac{p(E/Ho)}{p(E/Ho) + \frac{1-p(Ho)}{p(Ho)} p(E/not\ Ho)}.$$

reveals this conceptual duality, as well as the empirical convergence between the two probabilities; moreover, it defines the role of initial probability p(Ho). As the number of observations diverges, the asymptotic approximation of inverse probability to direct probability reduces the role of preliminary knowledge into inertial statistical aggregates, even though this information often remains relevant within the empirical limits in which experience takes place.

The theoretical confusion between p(E/Ho) and p(Ho/E) is what Corrado Gini called "the original sin of probability calculus". It tends to be perpetuated by the use, not empirically unfounded in the context of large numbers, of direct probability in place of inverse probability. Gini sometimes used the witty example of the policeman who, in the name of the law of probability, arrests a man who wins four numbers at lotto, under the accusation: "There was only one against five hundred thousand chances of winning by pure chance; there are thus five hundred thousand chances to one that the win was not due to chance". One would clearly smile at this line of argumentation. So, why do we not smile at the numerous inversions of inductive probability where the paralogism of the policeman is often present? Even more so when the probability is referred to a single event.

And it is precisely in the case of a single event that preliminary probability, i.e. the preconception, conditions induction. For example, what is the probability p(Ho/E) that the living phenomena (E) is due to a remote casual encounter between primordial inorganic constituents? Whatever value is given to "direct" probability p(E/Ho), "inverse" probability p(Ho/E) will depend on the value attributed to "a priori" probability p(Ho), i.e. on subjective convictions. The creationist has no doubt: he poses p(Ho)=0 and thus obtains p(Ho/E)=0. The materialist, on the other hand, gives p(Ho) a value equal (or near) to one, thus obtaining a completely different answer. It is therefore a problem of conceptual paradigms. There is the fact that the admissibility of a hypothesis regarding a single event can never be merely dependent on a probability. This is not the case if the event is a statistical one; it is therefore not the case of scientific research.

VII

Almost alien to scientific thought, utilitarian and subjectivistic conceptions lead in their extremes to an economistic transposition of inductive inference: probability as a convenient gamble, statistical induction as the choice of the most profitable solution. The encounter between Bruno de Finetti's subjectivism and Abraham Wald's decisionism gave rise to a transformation of statistics into a "managerial" form of mathematics, under the impulse of a utilitarian philosophy which regards an entire critical tradition as an immense prehistory.

Yet, it is not the same thing to class prevision as a problem of logical induction or one of optimization; and neither can a scientist's conjecture be put in the same class as a gambling bet. We must also not ignore the alien nature of strategic thought to a scientific mode of thought: the former seeks the most useful solution, the latter, the most truthful; the former minimizes a function of loss, the latter admits no form of penalization except for a back-tracking over hypotheses and evidence. Prevision may be a probable act, but this is no reason to reduce it to the terms of "Pascal's gamble" which proves it is always useful to believe in God: if he exists, to believe in him is highly advantageous; if he does not exist, nothing has been lost. Rather than a methodology of knowledge, this is an opportunist canon of a methodology of convenience in which utility has supremacy over truth, decision over induction.

Even should we choose to replace knowledge with decision, hypothesis with gamble, probable induction with strategic opportunity, we must ask ourselves whether it is the same to "play" with a nature which "chooses" once and for all or with a nature whose destiny is determined at the moment it is wrought, with no necessity in the emerging reality, it being just one of the many possible realities. After the "Origin of species" (1859), this become the intellectual (and moral) code of naturalistic knowledge, finding itself up against an evolving reality expressed by individual variability and the accidentality of the processes involved. "Chance" is redifined, on the one hand deriving from a lack of knowledge, on the other, from the non-deterministic interpretation of the occurrence of phenomena: it is the scientific world view which has made evolution the interpretative key of life processes and quantic determinism the universal grammar of elementary physical events.

Even in reducing a problem of inductive logic to one of optimization, we still have to adopt a canon with which to interpret phenomena as it is clearly not the same to bet on a necessary reality as it is to bet on a contingent reality: a reality which is, and might not have been. Once again, it is question of general hypotheses. We must draw a distinction between the uncertainty regarding an event, whose occurrence cannot be predicted exactly as not all the initial data are known, where probability is used due to a lack of information, and the state of uncertainty regarding a phenomenon, the occurrence of which is not implicit in the initial conditions, where unpredictability is immanent to the phenomenon itself.

Thus we return to our first question: which scientific environment offers a specific euristic reason to be for a methodology which reduces the proof of a hypothesis to a convenient choice? What natural philosophy provides for the transformation of knowledge into convenience, truth in utility? As the statistics produced by science is not the statistics for strategic interpretation, so the probability of the genetist or physicist is not that of the businessman or company consultant.

There is a categorical difference between the nomic value attributed to the probability of a quantic jump in an elementary particle, or DNA polymer nucleotid, and the economic value attributed to the probability of an optimal decision: rather than two different conceptions of the probability calculus, they are two distinct interpretations of the relationship between man and his world. They are two different ways of confronting the crisis of classic determinism, the predominance of an increasingly statistical view of reality: a reality which is variable, an immanence which becomes contingent.

University of Bologna (Italy)

L. Jonathan Cohen

SOME LOGICAL DISTINCTIONS EXPLOITED BY DIFFERING ANALYSES OF PASCALIAN PROBABILITY

At least six different kinds of theory about the semantics and epistemology of Pascalian probability are easily distinguishable in the history of the subject. One kind of theory evaluates probabilities with the help of a principle of indifference, a second by reference to relative frequencies, a third by natural propensities, a fourth by actual or appropriate strength of belief, a fifth by multi-valued logic and a sixth by ratios of logical ranges. What are we to make of this embarrassingly rich diversity of philosophical analyses?

Of course, not all philosophical analysts of probability have claimed that their own analysis is the only correct one. For example, though indifference theorists[1] and Bayesians[2] have often done this, Carnap[3] originally thought that probability could be conceived either as a logical relation or as a relative frequency and others, like Nagel[4] and Mackie[5] have been even more open-minded and have acknowledged the existence of four or five different concepts of Pascalian probability. But a tolerant pluralism leaves many questions unanswered. In particular, we need to understand how it is that several very different concepts of probability are possible.

It will not do to tell us here, as Mackie does, quoting with approval from J.S. Mill, that 'names creep on from subject to subject until all traces of a common meaning sometimes disappear'[6] unless there is historical evidence that some such gradual process of linguistic expansion accounts for all the differences under examination. But Mackie offers no historical evidence to support his thesis. Indeed, some diversity of conception was present at the outset. Even in the earliest years of work on the mathematical principles of Pascalian probability those principles were already being applied both to judgments of aleatory

E. Agazzi (ed.), Probability in the Sciences, 67–76.

chance and also to judgments of evidenced credibility. Compare judgments of goodness. Most philosophers would agree with Aristotle[7] that the various uses of the word 'good' in 'good man', 'good health', 'good pen', etc. are not due to chance homonymy. No more should we suppose that the various available senses of the word 'probable' (and of its equivalents in other languages) are due to accidents of linguistic history.

Nor is any light light cast on the problem by Mackie's claim[8] that probability should be seen as a 'family-resemblance' concept in Wittgenstein's sense[9], since that approach to the understanding of conceptual pluralism is inherently unrewarding. For example, men ride petrol-tankers, horses and tractors, but not oxen; oxen, horses and tractors, but not petrol-tankers, are used on farms for pulling things; petrol-tankers, tractors and oxen stay on the ground, while horses sometimes jump; and all petrol-tankers, oxen and horses have their front and rear means of locomotion approximately the same height, while most tractors do not. But no-one supposes it appropriate to mark this particular nexus of family resemblance by carrying over the name of one of the four sorts of objects to describe the other three sorts. The fact is that some groups of four or more sorts have common names, and some do not, even when the sorts exhibit family resemblance to one another. So unless the exponent of a family-resemblance approach to probability tells us why his supposed nexus of family-resemblance generates a common name, in contrast with others that do not, he has not explained anything. But if he does do this, and does it adequately, he has to go beyond a merely family-resemblance account. It will certainly not be adequate to say that a common name is generated when the resemblances are sufficiently close, if the only criteria of sufficient closeness is the use of a common name.

A deeper level of philosophical understanding is clearly necessary here. To achieve it we must begin by paying some attention to the fact that the different kinds of analysis of Pascalian probability that philosophers have proposed differ not only in the substance of their central message, but also in what they imply about certain important logical issues. Not all of these differences of implication have been adequately recognized in discussions of the relevant analyses. But when they are recognized they help us to see that behind the different kinds of philosophical analysis lie quite a variety of different conceptions of Pascalian probability that are available for human reasoning. For example, let us consider first the modality of probability-judgments, then whether they are sentence-related or predicate-related, then their substitutivity conditions, and then what may usefully be termed their 'counterfactualisability'.

Thus, if you adopt an indifference analysis for a particular judgment of probability, and assume the value of the probability to be determined ultimately by the rules of a game, that judgment is made out to be either necessarily true or necessarily false. If in a valid throw, a die can fall only on one or other of its six sides and if there is assumed to be an equal chance of its falling on any one side, there is no possible world in which, when such a game is played, the licit probability of its falling twice with the same side uppermost is anything other than $^{1}/_{36}$. Similarly Carnap's logical range theory imposes necessary truth or necessary falsehood on any evaluation of a particular probability-function for a particular pair of sentences, such as $c^{*}(h,e) = {}^{1}/_{3}$ or $c^{+}(h,e) = {}^{1}/_{2}$. But relative frequencies, natural propensities, or strengths of belief (or empirically established indifference-dependent chances) could be different in other logically possible worlds than the actual one. So judgments of probability in those terms can at best be contingently - not necessarily - true.

Again, we commonly take the overall structure of a probability-judgment to be represented by the symbolism $p(A) = n$ or $p(A/B) = n$. It is understood here that the letter n stands in for a real number, or for a rational one, such that $1 \geq n \geq 0$. But what do the letters *A* and *B* stand in for - whole sentences, expressing fully determinate propositions, or just sentence-schemata or parts of sentences? Carnap's analysis requires the letter *A* and *B* to stand in for sentences, since it is concerned with probability as a confirmatory relationship between propositions, and the multi-valued logic analysis also is obviously sentence-related rather than predicate-related. But the relative frequency theory requires *A* and *B* to stand in for terms designating classes, since it is concerned with the relative frequency with which members of one class are also members of another. And the indifference, propensity and personalist kinds of theory seem in principle to be compatible with both options, through particular versions of these analyses may not be. For example, you can speak in general about any game of drawing a card at random from a well-shuffled pack, and then '*A*' in 'p(*A*) = $^1/_3$' will signify a type of outcome, viz. the drawn card's being an ace; so '*A*' will stand in for a sentence-part. Or you can speak specifically of the very next round of playing that game and say that there is a probability of $^1/_3$ that the next next card drawn here this afternoon will be an ace, in which case *A* will stand in for a whole sentence. Again, you could measure both the strength of my belief, for any person, that he will survive to the age of 70, given that he's a lorry driver, and also the strength of my belief that our friend John Doe, specifically, will survive to the age of 70, given that he is a lorry-driver. In the former case '*A*' would stand in for a sentence-schema, in the latter for a fully determinate sentence.

Another important issue is that of substitutivity. Under what conditions of relationship between *A* and *A*' and between *B* and *B*' is the truth of p(*A*/*B*) = n sure to be unaffected if we replace *A* by *A*' and/or *B* by *B*'?

When the mathematics of Pascalian probability is treated as a multi-valued logic (and probabilities are consequently treated as truth-values), the logic is not truth-functional, because p(*A*&*B*) is dependent on p(*A*) and p(*B*/*A*), not on p(*A*) and p(*B*). So it is clear that propositions with the same truth-value can't always be safely substituted for one another within a judgment of probability.

In certain other analyses of probability further restrictions on substitutivity also operate. For example, Carnap's analysis makes p(*A*/*B*) depend on the ratio of the range of *A*&*B* to the range of *B*. This ratio is 1 where *B* logically implies A (because then the range of *A*&*B* is identical with the range of *B*). Now consider the Carnapian representation of the case where *B* is a conjunction of atomic sentences that logically implies *A*, and *B*' differs from *B* only insofar as one of the atomic conjuncts in *B*' has a predicate that differs from the corresponding predicate in *B* but is co-extensive with that predicate in its application within the actual world. Under these conditions the ratio of the range of

A&B' to the range of *B'* will not be 1, because *A&B* and *B'* will not be true in just the same possible worlds. So p(A/B) = 1 will be true, but p(A/B') = 1 will be false. For example, though it is certain that this man is George's sibling, given that he is George's brother, it may be only probable that the man is George's sibling, given that he is George's partner, event though it is in fact the case that all and only George's brothers are his partners. Thus only those predicative expressions that are analytically equivalent can be safely substituted for one another within Carnapian c-functions. And some restriction on substitutivity, albeit a lesser one, applies also to a propensity theory, because accidentally co-extensive predicates do not attribute the same property or natural tendency. For example there may be a strong natural tendency - equalling a probability of, say, .8 - for persons elected to Parliament to have been educated at a university. But even if it were accidentally the case, through a surprising set of coincidences, that all and only Parliamentarians turn out to have names with eight vowels we should not suppose this feature of people's names to *depend* to any extent on their having been to a university: the probability that such a person had been to a university would be a relative frequency, not a propensity. For predicative expressions to be safely substitutable for one another within a statement of a propensity-type probability, their equivalence must be guaranteed by such laws as those of nature, logic, mathematics or language.

Other kinds of substitution are less restricted.

For example, on a relative frequency account it obviously makes no difference to the value of a probability if we use a different expression to designate the same class. If the passengers on a particular train have return tickets if and only if they intend to return the same day, the the relative frequency of passengers with return tickets is just the same as that of passengers intending to return the same day. So on a relative frequency account co-extensive predicates - predicates satisfied by all and only the same individuals - can be substituted for one another within a judgment of probability without any risk of altering the judgment's truth-value. And the same holds good for the realist form of indifferentist account, since the actual ratio of favourable to equally possible cases should be quite unaffected by a change in the expression used to designate the outcome of which the probability is at issue.

Even on a personalist analysis probability-judgments are by no means as resistant to substitution as personalist talk about probability's measuring strength of belief might lead one to suppose. In everyday descriptions of people's beliefs not even logically or mathematically equivalent expressions may safely be substituted for one another, let alone expressions that refer to the same individuals or predicate the same properties[10]. And the same holds good for probability-judgments if we accept a personalist analysis that does not insist on what de Finetti called coherence. But, where coherence is required, the restriction on substitutivity has to be discarded: the context needs to be referentially transparent. Anyone whose probability-judgments are to be taken as declarations of the lowest odds that he is prepared to accept will guard himself adequately against a Dutch book only if he treats all his ordinary probability-judgments as being open to the mutual substitution of terms that have the same reference or meaning as one another. You can't allow yourself to lose a bet just because the person you bet with uses a different designation for the same outcome: a horse runs just as fast (like a rose smells just as sweet) under another name. The only wagers in which such substitutivity is restricted are those on the truth of propositions that are resistant to substitution even when they are not the subject of a wager: for example, propositions about a person's fears, hopes, desires, beliefs or other mental states.

Important questions of substitutivity may also arise even where the two expressions that are candidates for intersubstitution do not designate the same thing. Consider a monadic or dyadic judgment about a specified individual, such as John Doe (as distinct from an explicitly general judgment about 'any' person). We might say, for example, that the probability of John Doe's surviving to age 70, given that he's a lorry driver, is .8. Now does this statement refer to John Doe as just one individual among many others, so that if the probability holds for John Doe it must hold also for Richard Roe or for anyone else? Or does sit refer to John Doe in all his singularity of circumstances (healthy exerciser, cautious eater, non-smoker, etc.), so that the probability of Richard Roe's survival to age 70, given that he's a lorry driver, may be quite different? In the former case the name 'John Doe' may be replaced by any other expression that designates a particular individual, in the latter it may not. So let us use the term 'implicitly general' to describe those grammatically singular judgments of probability about a specified individual which necessarily remain true whenever expressions denoting that individual are uniformly replaced by expressions denoting any other individual within the same domain of discourse. And let us use 'implicitly singular' to describe a grammatically singular judgment of probability in which such an expression is not necessarily replaceable without affecting the judgment's truth-value. Then an indifference theory will impute implicit generality to grammatically singular judgments of probability. Also singular judgments made with the use of any of Carnap's symmetrical confirmation-functions will be implicitly general, though singular judgments made with the use of non-symmetrical confirmation-functions will be implicitly singular. Again, a propensity theory will require grammatically singular judgments of probability to be implicitly general, insofar as the characteristic evidence for claiming the existence of a conditional propensity is an observed relative frequency. The observed ratio of *A*s among *B*s is just as good evidence for the strength of the tendency for any one particular *B* to be an *A* as it is for the strength of any other *B*'s tendency to be an *A*. But a personalist theory need not require grammatically singular judgments of probability to be implicitly general unless the theory is confined to dealing with what de Finetti calls 'exchangeable' events. Certainly my belief that John Doe will survive to the age of 70 may well be stronger than my belief that Richard Roe will, and the lowest odds that I would accept on the one outcome might well be lower than the lowest that I would accept on the other, just as I might expect often to bet at different odds on different horses in the same race. So on a personalist analysis grammatically singular judgments of probability are implicitly singular also, unless otherwise stated[11].

Yet another important logical issue concerns what is best called 'counterfactualisability'. This issue arises only in regard to those judgments of probability that are both dyadic and implicitly or explicitly general. Such judgments typically concern the probability that an entity is *A* given that it is *B*. But does such a judgment apply even to entities that are not actually *B*, in which case it may be said to be 'counterfactualisable', or only to entities that are actually *B*, in which case it may be said to be 'non-counterfactualisable'? For example, to accept that, on the assumption that he is a 40-year-old asbestos worker, a man has a .8 probability of death before the age of 60, is to accept an explicitly general conditional probability that applies not only to any person in the history of the universe who is actually a 40-year-old asbestos worker but also to any individual who is not. That is, the same judgment of probability would presumably hold good even if the number of asbestos workers were larger than it in fact is. Only thus might someone legitimately infer a reason for not working in an asbestos factory. But to accept a .6 probability, for any person picked out at random at a certain conference, that he is staying in the Hotel Excelsior is not to accept a probability that could be relied on to have held good if more people had been at the conference. Perhaps the additional participants would all have had to stay elsewhere because the Hotel Excelsior had no more unoccupied rooms. Such a probability-judgment states an accidental fact and is non-counterfactualisable, whereas the judgment about mortality among asbestos workers reflects a causal connexion and is therefore counterfactualisable.

Just as some universal truths are counterfactualisable and some are not - compare 'Everyone who drinks cyanide is poisoned by it', with 'Everyone at this session is staying in the Hotel Excelsior' - so too some probabilist-judgments are counterfactualisable and some are not. But (in regard to probability-judgments that are both dyadic and general) a propensity theory seems applicable only to counterfactualisable judgments. If a natural tendency exists for a certain kind of event to occur in such-and such circumstances, it would normally still have existed even if those circumstances had occurred more often than they actually occurred. Similarly a relative frequency analysis imputes counterfactualisability where the reference-class has an infinite number of members, since any relevant convergence to a limit might be expected to maintain itself even if additional members joined the reference class. And an interval-valued relative frequency analysis also imputes counterfactualisability where the reference-class has an indeterminately large number of members, since the impact on any additional members sight be expected to be accommodated within the specified interval. But where probability is conceived of as point-valued and the reference-class is of finite size, a relative frequency analysis does not impute counterfactualisability since in that case the probability-judgment is normally supposed just to report actual ratios of occurrence. Similarly the indifference theory does not impute counterfactualisability, because it supposes a probability-judgments to state the ratio of the number of outcome-types of a certain specified kind to the total number of mutually co-ordinate outcome-types and to say nothing about what another outcome-type might have been like if it had been an additional co-ordinate outcome-type over and above the actual totality of them.

Thus (though the point is often unnoticed) different theories about the analysis of probability may in effect attribute different logical properties to probability-judgments. And we have to bear this in mind when we consider the relevance of different analyses to the various different purposes for which probability-judgments are required.

The Queen's College, Oxford University

NOTES

1 Laplace, P.S. de, (1951), p.4ff.
2 Finetti, B. de, (1964), esp. pp. 111-118
3 Carnap, R. (1950), pp. 23-36
4 Nagel, E. (1958), pp. 342-422
5 Mackie, J.L. (1973), pp. 154-236
6 Mackie, J.L. (1973), p. 155, quoting from Mill, J.S. (1896), p.24
7 Nicomachaean Ethics 1096b, 26-27
8 Mackie, J.L. (1973), p.155
9 Wittgenstein, L. (1953), pp. 32-33

10 Carnap, R. (1947), pp. 59-64; and Quine, W.V.O. (1960), pp. 143-147

11 Pace Carnap, who claimed that 'all authors on probability, both classical and modern' must be taken to have accepted the implicit generality of these judgments. Cf Carnap, R. (1950), p. 488

BIBLIOGRAPHY

Aristotel, Nicomachaean Ethics

Carnap, R.:

(1947), Meaning and Necessity. Chicago: University of Chicago Press.

(1950), Logical Foundations of Probability. Chicago: University of Chicago Press.

Finetti, B. de, (1964), 'Forsight: its Logical Laws, its Subjective Sources (1937)'. In Studies in Subjective Probability, edited by Kyburg, H.E. and Smokler, H.E.. New York: John Wiley and Son.

Laplace, P.S. de, (1951) A Philosophical Essay on Probabilities. New York: Dover.

Mackie, J.L. (1973), Truth, Probability and Paradox. Oxford: Clarendon Press.

Mill, J.S. (1896), System of Logic.

Nagel, E. (1938), "Principles of the Theory of Probability". In Foundations of the Unity of Sciences, edited by Neurath, O., Carnap, R., Morris, C.. Chicago: University of Chicago Press.

Quine, W.V.O. (1960), Word and Object. Cambridge: Massachusetts Institute of Technology Press.

Wittgenstein, L. (1953), Philosophical Investigation. Oxford: Blackwell.

Craig Dilworth

PROBABILITY AND CONFIRMATION

1. That qualitative probability is as much a form of probability as is quantitative probability; and in what the difference lies. In considering the nature of probability, and its possible uses in and applications to the sciences, it is to be kept in mind that a fundamental distinction exists between what may be called the qualitative and quantitative notions of probability. The qualitative notion of probability may be associated with the simple adverbial and predicative expressions 'probably' and 'is probable', while the quantitative notion involves the assignment of a *measure*. While the qualitative notion may be reformulated in quantitative terms, namely as a probability greater than zero (is possible) or 1/2 (is probable), its normal employment does not commit the user to being able to specify the quantitative nature of the relation between the evidential basis of the probability claim and the claim itself. With the use of the quantitative notion, on the other hand, the user may be viewed as being so committed.

2. That probabilities are of something's being the case. Quite generally, we can say that a probability is invariably a probability of something's being the case. A similar view has been expressed to the effect that probabilities are paradigmatically, if not always, probabilities of (the occurrence of) *events*. But this latter view seems too narrow, and does not do justice to a variety of instances in which there is an agreement in our intuitions as to the viability of employing probability notions, especially qualitative ones. Thus, for example, we should have little hesitation in applying the notion of probability to the case involving quasars, saying that quasars are probably of the order of physical magnitude of galaxies, considering the evidence we have for the great quantities of energy they release; and here

E. Agazzi (ed.), Probability in the Sciences, 77–87.

we are not speaking of an event, at least not in any straightforward sense, but of something's being the case.

3. That the notion of probability can be applied to past and present states of affairs, and not only to future ones. Another delimitation of probability notions which seems too strict is the suggestion that they are to have relation only to *future* events or states of affairs. But can I not say of Socrates that he probably knew of Democritus; or, given adequate geological evidence, that the probability of a particular volcano's erupting during the palaeolithic era was such-and-such? Similarly, I should be able to apply probability notions to certain present states of affairs, such as that involving quasars above, of which I do not have complete knowledge. In fact, it is the notion of complete or sufficient knowledge that is key in this regard, and it is precisely in those cases where such knowledge is lacking that the concept of probability finds its application.

4. That we can speak of the probability of entities which have truth-values, if by this we mean the probability of their being true. A third delimitation of probability notions that does not seem entirely warranted is the denial of their applicability to statements, hypotheses, conjectures, guesses and other locutions to which we would normally attribute truth-values. For such locutions describe states of affairs, and, given that what we mean by a statement's being probable or having a particular probability is that it is probably true or that the probability of its being true is such-and-such, then we are only expressing in different terms the probability of what the statement describes being the case. And this is quite in keeping with what we expect of probability notions.

5. On two senses of the word 'confirm'. By allowing that we can speak of the probability of statements or hypotheses (being true), we bring the notion of probability closer to that of confirmation, for we can also apply the latter notion to such locutions. But here we must distinguish two senses of the word 'confirm': 'To corroborate, or add support to (a statement, etc.),' and 'to make certain, verify, put beyond doubt' (Oxford English

Dictionary). It is clear that if we are to align probability with confirmation, we must take 'confirm' in the first sense, which does not imply certainty. As will be suggested below, however, when one speaks of the confirmation of empirical laws, it is the second sense of the term that is to be applied.

6. *On distinguishing the evidential basis of a probability claim from its subject matter; and the relevance of this distinction to the statistical interpretation of probability.* In considering the nature of probability claims it is important to distinguish the *evidential basis* of such a claim from its *subject matter*, or what it is *about*. (On this point, cf. Harré, 1970, pp. 160ff.) The former often manifests itself in the form of particular data, of which we have knowledge. The latter, on the other hand, consists of the state of affairs (those states of affairs) regarding which our knowledge is lacking, and about which we are using the data to gain information.

This distinction is of particular importance with regard to the statistical or frequency interpretation of probability. It is one thing to say that (quantitative) probability is nothing other than what can be obtained on the basis of an empirical investigation of relative frequencies, and another to say that what a probability claim made on such a basis is *about* is itself relative frequencies. This distinction has been overlooked not only by critics of the statistical interpretation, but also by its supporters. Even von Mises misses it, where he is led to say: 'The phrase 'probability of death,' when it refers to a single person, has no meaning at all for us.' (p. 165 in Kneale). There is nothing in principle to prevent one's taking the evidential basis of a probability claim to be the relative frequency of the occurrence of events in an empirical sample, and its subject matter to be a unique state of affair.

7. *On methodological grounds, and their relevance to inductive probability.* As suggested in §1, the making of acceptable quantitative probability determinations requires there being a specifiable quantitative link between the evidence for the determination and the value determined. The nature of this link is dictated by the methodological foundation of the probability claim,

or the *rules* in accordance with which the claim is made. These 'methodological grounds' tell us first what is to *count* as evidence, and second how to obtain a probability value on the basis of that evidence -a procedure which includes a weighting of the evidence. All probability determinations are based on evidence and require methodological grounds, whether those grounds be implicit or explicit.

It is thus with some scepticism that we regard Carnap's suggestion that what he calls *inductive* or *logical* probability is on a par with yet distinct from statistical probability, and 'occurs in contexts of another kind.' According to Carnap, a statement of inductive probability is to be such that: 'If hypothesis and evidence are given, the probability can be determined by logical analysis and mathematical calculation.' (p. 271). This requirement, however, in that it asserts that there must be a clearly delimitable relation between the evidence for a probability claim and the value asserted by that claim, is simply the demand that (part of) the methodological grounds of such claims be specifiable. Though there is nothing to prevent Carnap's developing his view so as to provide new methodological grounds for probability determinations, what he here terms 'inductive probability' is but an aspect of any acceptable determination of quantitative probability.

8. On the implications of a probability claim, and the relevance of this notion to the subjective interpretation of probability. In §6 two aspects of probability claims were discussed: their evidential basis and their subject matter. Another aspect is what they *imply*, or, in certain contexts, what they *mean*. If I say that it will probably rain tomorrow, the evidential basis of my claim may be today's weather report; and its subject matter is tomorrow's weather. But the claim might *imply*, among other things, something about my state of mind, which in turn might set limits on what may be considered rational behaviour on my part. Thus my claim might be taken to indicate that I believe that it will rain tomorrow - the claim's 'illocutionary force'; and, assuming I believe this, that it would be foolish of me to begin preparations for a day at the beach.

The notion of what probability claims imply is of particular relevance to the subjective interpretation of probability, according to which probability determinations represent degrees of belief. Now if it is intended by this that degrees of belief are primary, then this view is quite unacceptable; for not only is there no way of directly quantifying the 'degree' of a person's belief, but different people can have different beliefs on the same evidence, and there would be no sense in speaking of certain of those beliefs as more correct than others. Each person would have his own opinion, and that would be the end of the matter.

The same argument applies to taking one or more persons' degree(s) of belief as the *evidential basis* of a probability claim. We would still lack a *reason* for their believing as they do.

A third line might be to say that what probability claims are *about* are degrees of belief. But, as is suggested by the above example, this is only to confuse the subject matter of a probability claim with its implications or illocutionary force. While my saying that it will probably rain tomorrow might be construed as implying that I believe to a degree greater than 0.5 that it will rain tomorrow, my claim is not about my belief, but, as mentioned above, about tomorrow's weather.

As a consequence of these considerations, we are forced to conclude that subjective probability, if not basically misconceived, is only of peripheral relevance to the theory of probability.

9. That the propensity interpretation of probability denies that probability is knowledge-relative; and that it is unscientific. Another interpretation of probability is the propensity or 'objective' view, according to which probability is a *property* of a state of affairs. With regard to the term 'objective' as used here, we note that it is (or should be) intended in the sense of 'residing in the object'; and not in the sense that would suggest this approach to be necessarily more objective than other approaches when it comes to the *method* by which probability assignments are made.

An important implication of the propensity view is that probabilities, in being objective properties of states of affairs, are not knowledge-dependent. Thus the probability of a particular die's turning up an ace being 1/6 is a property of the situation in which the die is cast, and is not dependent on our knowledge of that situation. But is this so? Let us say that the die is cast and turns up a five. This does not refute the objectivist's thesis. But let the die be cast again in a situation that is qualitatively *identical* to the former one. According to common sense and a basic principle of physics, since the situation is qualitatively *exactly* the same, the die will again turn up a five. The reason we speak of probabilities in the case of dice throwing is that we do not *know* precisely what the situation is in which a die is cast. The probability of obtaining an ace is not a property of the situation, but a function of our knowledge of the situation; and, following Laplace, we shall here generalize this and suggest that probability is *always* knowledge-relative.

The propensity interpretation has counter-productive effects when applied to science, for it condones ignorance. If the probability of something's happening is simply taken to be a property of the state of affairs in which it happens, then once one has probabilistic information there is no impetus toward a more detailed investigation of the state of affairs. One can, with equanimity, say that that is just the way things are, and no compulsion need be felt to attempt to discover more rigorous laws according to which the events in question occur. And such intellectual complacency, needless to say, is clearly unscientific.

10. On the difference between empirical and theoretical probability determinations; and its relevance to the distinction between a priori and a posteriori probability: the causal interpretation of probability. When a particular probability distribution is empirically determined always to obtain given conditions of a certain kind, we are led to conclude, as suggested by Laplace, 'that the ratio is due, not to hazard, but to a regular cause.' (p. 393 in Harré, 1967). Were there not such a cause in operation, we should not understand why just that probability distribution is manifest given conditions of that kind.

In some cases the cause may be relatively evident, as when our concern is with the casting of dice; in others it may be hidden. When it is hidden, we, as scientists, may *theorize* about its nature. In the best of cases we can construct an acceptable theory which enable us empirically to recognize the difference in the conditions under which each of the elements in the distribution is manifest, thus allowing us to replace probability with certainty. But even if our theory is unable to do this, it might nevertheless explain why the distribution as a whole takes the form that it does. Thus, for example, while on the empirical level we can see that a certain proportion of the children of parents with blue eyes and brown eyes themselves have e.g. blue eyes, by moving to the level of the genetic theory we come to understand the cause of this phenomenon. And, given that we are warranted in claiming that our knowledge has been extended to the nature and functioning of genes, we might go so far as to say that the probability value we have obtained from our empirical investigation of a (necessarily) finite number of cases is not quite the correct one, and furthermore what the value should be if based on an indefinitely large sample.

Here, then, even assuming the probability value obtained to be the same in the two cases, the *evidential basis* of the latter determination would differ from that of the former. While evidence for the empirical determination would also be evidence for (or against) the theoretical, if we claim actually to know that the theoretical mechanism exists, evidence directly affirming its existence over and above its ability to produce probability values near to the empirical ones would also constitute evidence for the viability of the values it suggests. Also, the *methodological grounds* on which the theoretical probability assignment is based, in that they concern how the actions of the theoretical mechanism should be manifest empirically, would also differ from the rules employed in the purely empirical case. This brings us to a consideration of current usage of the terms 'a priori' and 'a posteriori' in probability theory. It is here suggesteds that probability determinations of the 'theoretical' sort constitute an instance of what can justifiably be considered determinations of *a priori* probability, while those of the empirical sort are *a posteriori*. This distinction can also be applied to simpler cases. thus we should say that an unloaded die, cast under normal conditions, has the a priori probability 1/6 of turning up an ace, or that the a priori probability of getting zero on a fair roulette wheel is 1/37, and contrast this with the a posteriori probability of such occurrences based on observed frequencies, the value of which may be different. (On this point, see Kneale, §37.) Recognition of this distinction also leads us to say, with Kneale, that to deny or ignore the viability of probability determinations of the sort here termed a priori - as is done on the statistical interpretation- is but a form of positivism.

The conception of probability sketched above may be called the *causal* interpretation; and it is the view supported in the present essay.

11. That for methodological reasons probability values cannot be determined for locutions that do not have truth-values. In order to make a quantitative probability determination, whether a posteriori or a priori, not only must we be able to put the methodological grounds of the determination in a quantitative form, but we must also have a clear concept of its subject matter. For example, if we say that the probability of a particular coin's turning up heads on the next flip is 1/2, we must have a clear notion of what it means for the coin to turn up heads.

This point is not problematic in games of chance, where the occurrences with regard to which probability calculations may be made are clearly delimited. But when we move over to considering states of affairs in a broader context, this clarity may be lacking. Thus, for example, we cannot speak of the probability of the cat's being on the mat, if we are not clear as to whether for this to be the case the cat's tail must also be on the mat. In other words, the application to the cat of the predicate 'being on the mat' must obey the law of the excluded middle: i.e. the statement that the cat is on the mat must have a truth-value, in the sense that under any conditions it must be either true or false.

12. On the acceptability of scientific theories. Considerations to this point suggest that we would face insurmountable difficulties should we attempt to attribute probability values -whether or not under the name of 'degrees of confirmation'- to the various expressions of the natural sciences. In the case of scientific *theories*, each of which postulates the existence of a hitherto unknown realm of entities whose behavior is to be responsible for certain empirical regularities, we must first determine what is to *count* as evidence for or against the theory in question.

As has been pointed out by Harré (1970, p. 166), this involves a consideration of more than simply how the theory should manifest itself empirically, for we also have to take into account or treat as evidence the reasonableness of assuming the mechanism the theory posits as actually existing and being responsible for the regularity. Bohr's theory of the atom, for example, can well account for the spectral lines of gases; but in so doing it suggests that electrons within the atom move from one orbit to another without occupying the intervening space. This contravenes a fundamental scientific principle, and its doing so cannot be ignored when we come to consider the likelihood of what the theory suggests actually being the case.

But even if we recognize this fact and say, for example, that one of two empirically equivalent theories is more likely to be true if it does not contravene such principles, we still do not know how to weight the two kinds of evidence, particularly with respect to each other. Their difference seems to be fundamentally of a qualitative sort, and it is virtually impossible to imagine how one might arrive at a weighting procedure which is not highly arbitrary.

Also, due to the idealizational nature of scientific theories, it is not at all clear what would have to be the case for us to say of them that they are *true*. The kinetic theory of gases treats molecules as though they were perfectly spherical. Is anything in reality perfectly spherical? Taking the idealizational nature of theories into account we would simply have to say, granting that they have truth-values, that all theories are false; and as a consequence it would be pointless to attempt to apply probability values to them.

What is done instead, as is reasonable, is that one speaks of the *acceptability* of theories, either alone or as compared to other theories. And it is not imagined that this acceptability comes in degrees computable with the aid of an algorithm, but that it is qualitative and determined by the informed judgment of scientists.

13. On the confirmation of experimental laws. It may be thought, however, that while we may not be able to speak of the quantitative probability (or degree of confirmation) of scientific *theories*, we still might be able to speak of the quantitative probability of empirical *laws*.

Here again, however, we face the problem of determining what is to count as the entity in question being true. No empirical laws are strictly true, or hold exactly, but all are relatively applicable within a certain range of experimental conditions, and less applicable outside that range. Thus if we admit that such laws are not entirely false, the requirement stressed in §11, that there must be no vagueness as regards the statement with regard to which the probability assignment is made, is not met in this case.

Another aspect of empirical or experimental laws that deserves mention here is that they are determined in accordance with certain scientific principles, the most notable being that of the uniformity of nature. It is for this reason that an experiment need be repeated only a few times, mainly to ensure that no mistake has been made in the way it has been conducted, in order for its results to be established and incorporated into the body of scientific fact. Thus, though the (qualitative) probability of the results of the experiment being correct increases when the results are repeatedly obtained by re-staging the experiment, after only a few such re-stagings these results are accepted as certain, and no longer as merely probable. And it is in this sense that one speaks of empirical laws as being *confirmed*.

14. On the applicability of scientific principles. Such scientific principles as that of the uniformity of nature, which are usually tacit, and the conservation of energy, which are explicit, set guidelines for the conducting of scientific enquiry, both empirical and theoretical. That they should be considered either true or false is also problematic, for they function more as *rules* for the doing of science. Though it is admitted that they may face counter-instances, in such cases we are more inclined to question whether we are *applying* them correctly, and do not claim that they have been shown to be false. But even if we were to speak in such terms, it is impossible to point to real states of affairs which show them necessarily to *be* false, for we can never be sure that we know everything of relevance to the situation in question that would warrant our drawing such a conclusion. Thus scientific principles are seldom discarded entirely, but are more often set aside in those situations to which they do not seem to apply.

As a consequence of this, if we consider applying the notion of quantitative probability or confirmation to such principles, we face difficulties at every turn. We cannot say that what apparently constitutes evidence for or against a particular principle necessarily does so, for our lack of total knowledge of the relevant situation may mislead us on that score. And so we cannot give a numerical weight to the evidence, nor can we indicate the methodological grounds which tell us how to assign a probability value to the principle itself.

Consideration of the actual natures of scientific laws, theories and principles suggests not only that a project aiming to determine their quantitative probability or 'degree of confirmation' could never be realized, but that the very idea of providing an algorithm for making such determinations is based on an overly simple conception of the nature of science.

Abo Akademy (Finland)

REFERENCES

Carnap, R. (1955), 'Statistical and Inductive Probability,'. In The Structure of Scientific Thought, edited by Madden, E.H.. Boston: Houghton Mifflin Co.

Harré, R. (1967), 'Pierre Simon de Laplace,'. In vol. 4 of The Encyclopedia of Philosophy, edited by Paul Edwards. New York: Collier Macmillan Pub.

Harré, R. (1970), The Principles of Scientific Thinking. Chicago: University of Chicago Press.

Kneale, W. (1949), Probability and Induction. Oxford: Clarendon Press.

Giuseppe Del Re

CHANCE, CAUSE, AND THE STATE-SPACE APPROACH.

1. INTRODUCTION

As is well known, the concepts of chance and cause were the object of profound studies by most great scientists and philosophers of all times, from Aristotle to Kant, from Poincaré to Bornand to Prigogine. Yet, in the praxis of scientific research, we still tend nowadays to take for granted a classical, deterministic notion of scientificity which does not take into account the major innovations in the conceptual heritage of science brought about by statistical physics, quantum mechanics, the theory of evolution, information theory, and computer science. As has been pointed out by Bunge (1987), it would be of special interest to try to reformulate the study of causality (and, as a consequence, of randomness) within the conceptual frame of the "state space". Although this note is mainly devoted to chance, which is a concept much less discussed than causality, it is also intended as a contribution towards that aim, inasmuch as it tries to indicate how the transition from traditional language to state-space language arises from a discussion of chance within the natural sciences, it provides an analogy with the now familiar workings of a computer, and finally tries to hint at the limitations of the state- space conceptual frame with respect to metaphysical issues.

Since this note reviews a field already explored by countless outstanding philosophers, detailed references would occupy more than the text. Therefore, we have made no attempt to provide them, and only a few recent books which are especially relevant to the whole topic and contain an extensive bibliography will be referred to.

E. Agazzi (ed.), Probability in the Sciences, 89–101.

We must start, of course, from an attempt to formulate a satisfactory definition of chance, and of related philosophical and scientific notions, such as probability. It is somewhat paradoxical that the former iseasier to define than the latter, and yet is a psychologically indigestible concept. While probability is a notion belonging to everyday life, real chance is never accepted, and our ignorance of the laws governing the world or of the will of some god, such as Fate, are normally assumed to be the reason for apparently random events. This psychological difficulty stems form the deep relation that exists between the metaphysical and the physical notion of cause, between randomness and contingency.

Since chance is essentially related to the natural sciences, we shall use here the term 'cause' within the context of science, i.e. as a concept that can be described in terms of space, time and matter, and fits a theory where the relations between phenomena are mapped onto logical relations between statements concerning entities and events in spacetime. As a tentative interpretation of what is meant by cause in the current practice of science - to be revised in the following - we take the statement that a physical cause is some feature of the state of the universe at a given instant such that it is a necessary and sufficient condition for the next state to have certain characteristics - or for a specific event to take place.

We call attention to the concept of universe just introduced, which is essential (and, as far as we know, not sufficiently emphasized) to studies such as ours. As will be clearer from its use in the following, here that concept is indeed akin to the concept of an isolated system in physics, inasmuch as it treats as non-existent all other physical systems, but also has a philosophical connotation, for it also treats as non-existent agents, events or entities that do not fall within the given conceptual approach, in our case the observational-logical approach of natural sciences.

2. A DEFINITION OF CHANCE IN NATURAL SCIENCES.

We shall now take as a definition of chance the following negative one: chance is absence of (physical) causes. This definition can be illustrated first of all by Aristotle's well known analysis of chance (Aristotle 1961). Aristotle saw chance as the encounter of two independent causal chains, deploying themselves along space-time lines belonging to two separate regions of the universe. The fortuitous event of the creditor meeting the debtor because both had independently decided to visit the cattle market illustrates precisely this point, provided we ignore here free will, and take the decisions as belonging to some deterministic mechanism. Then, the presence of the two persons at the market is fully determined for each, the random event is their occupying the same point of space at the same time.

It is important to emphasize that in this example neither of the two personages could have foreseen the encounter, but a third person, knowing the 'history' of both, could. In 'objective' language, this can be stated by saying that the fortuitous event exemplified by Aristotle is such only if the two causal chains never had any common point in the past. It may well be that that the total independence of the two causal chains only holds because of a limitation of our 'universe' to a specific class of facts, in our case to phenomena susceptible of a scientific description. This means that we can have relative as well as absolute chance. We speak of the latter only if -using the terminology of Salmon (1984)- the universe formed by the all physical systems admits the possibility of processes not belonging to causal forks. And this still leaves open the possibility that causal forks always exist if the limitation to the natural sciences is also removed. Before proceeding further on the implications of the above definition, we must also deal with one major objection. Consider the case of a man injured by a roof tile while walking on a sidewalk. We normally say that the injuries were an accident due to chance, and yet there is a cause, the impact of the tile. The answer to this objection is that the random event where chance is called into play is not the injury, but the <meeting> of the tile and of the head. That event as such has no cause, even if different chains of causes have independently caused the tile and the head to occupy the same region of space at the same time.

The above counterargument is based on the assumption that the injury and the meeting are different events, i.e. are separable in principle in space-time. This point is specific of physical causes, and was emphasized already by Kant (1956):

> The time between the application of a cause and its immediate effect may be a vanishing quantity, but the relation of the one to the other will always remain determinable in time. If I view as a cause a ball which impresses a hollow as it lies on a stuffed cushion, that cause is simultaneous with the effect. But I still distinguish the two through the time relation of their dynamical effect.

3. CHAOS AS A REFERENCE STATE.

The definition given is implicit in a fundamental remark of the great physicist Max Born (1951):

> I think chance is a more fundamental conception than causality, for whether in a concrete case a cause-effect relation holds or not can only be judged by applying the laws of chance to the observation.

Chance, in other words, is the fundamental reference concept of physics, and the so-called law of chance is the fundamental law of physics, inasmuch as the existence of a (physical) law proper - another type of regularity or a causal law - is revealed by the fact that the normal Gaussian distribution is not applicable. Born's statement also emphasizes that physical laws are not necessarily 'causal' laws even in a very weak sense.

Of course, the results of a repeated experiment may not obey the normal distribution, and yet not be entirely determined by a causal law. In such cases, the events under study are not governed by necessity, and yet their randomness must be considered only partial, in agreement with the opinion that the normal Gaussian distribution is only one extreme of a whole range, whose other end is complete determination.

Indeed, in most observed events there is only partial randomness, and the extent to which they are predetermined by the state of the universe immediately preceding them is measured by probability. Whether in such cases we should speak here of a statistical law or of a causal law insufficient to determine completely the realisation of the observed event is a question which we shall not discuss here (cf. Del Re 1984).

4. PREDICTABILITY AND REGULARITY.

Strictly speaking, unless one refers to the absence of intentionality, randomness as such implies unpredictability. The converse is not true. We do not say that a newborn baby weighed four kilograms by chance, nor do we say that the baby was born on a Monday by chance, because we feel that there must have been some specific ground (the individual rythms of the mother, for instance) for such details. Thus, a statistical law (such as the constancy of the average weight at birth or the duration of pregnancy) is possibly a partial law, but does not imply that events governed by that law are incompletely determined.

The above considerations suggest that we have to distinguish between a regularity and a causal law, between predictability and determination, as has been discussed extensively both by Bunge (1987) and by Salmon (1984). We have partial or total randomness when the previous history of the system under consideration (the "universe") does not contain all the necessary and sufficient conditions for the event to take place.

It must be noted explicitly that this has nothing to do with the characteristics with which the event will present itself, if it takes place at all: there may be laws governing such characteristics, which are not causal laws, for they do not take the form: "given the state A of the universe, the state B follows necessarily". They rather take the form: if the universe is in state A at a certain time, the state B immediately following it must have certain characteristics - must be 'compatible' with A -, but will be one of a set. In the simplest case we have the alternative A/B, which means that the existing laws determine all the characteristics of the new state if it is realised, but cannot tell whether it will be realised or not. For example, in a collision between two atoms, it is possible to predict the nature of the diatomic molecule which may be formed, but not whether it will actually be formed.

Causal laws are thus special inasmuch as they tell something not only about the characteristics of the state B of the universe which follows the state A observed at a given instant, but - possibly with due account of the whole history of the universe up to the given instant of time - imply that the new state B is already contained in state A. Indeed, we can say that such a law really amounts to two distinct laws: one assigning the predicates of the state-to-be, the other stating that it will inevitably follow the present one. Since the latter kind of statement, from the point of view of natural science, can be subjected to the well known critique of Hume, we may conclude that the very concept of cause should be avoided, in agreement with a remark by Lalande (1968):

On peut observer en effet que la notion de cause dans les sciences est d'un usage d'autant plus rare qu'elles sont plus developpées, et qu'elle tend à être remplacée par des lois énonçant la permanence ou l'équivalence de certaines grandeurs.

5. THE ANALOGY WITH A COMPUTER.

The purport of the problem at hand may be further analysed by recourse to the analogy between the given 'universe' and a computer executing a programme. As is well known, a computer has an internal clock which will force it to pass automatically from one state to another if the '*instruction*' specifying the operation to be performed in order to reach the next state is found. This means that the mechanism by which the computer goes to next state is distinct from that by which it selects it. An ordinary instruction specifies completely the next state (=the set of sequences of digits 0 and 1 assigned to the available memory locations). It is easy, however, to push the analogy to cover also the case of incompletely determined events. In fact, suppose that at a certain moment the machine reaches instruction #20, which reads (X being an arbitrary integer):

```
20 Y = RDN (X): R = 100*(Y+10): GOTO R.
```

Since RND(X) specifies a random integer between 0 and X, the computer will branch out at random to one of the instructions 1000, 1100, 1200, 1300, 1400, 1500, 1600, 1700, etc.: we are thus simulating a process which obeys a law *analogous to* quantum mechanical uncertainty and to the interplay of chance and necessity postulated by theories of biological evolution. In order to cover the latter case, the analogy can be completed by adding to the program an "IF" instruction simulating a selection rule. For instance, it may be stipulated that any one of the instructions labelled with a possible value of R computes a specific real number N, and then ends by a GOTO S. Instruction S will be an "IF" instruction, such that certain values of N are accepted for printing, others are not. This can be interpreted as representing the fact that, whenever Y or N do not satisfy the test, the 'mutation' represented by the N-value computed in instruction 100*(Y+10) is refused, and another random choice for Y is requested. The following program is an extremely simple version of the above scheme working with 8 states (X=7; S=2000):

```
20 RANDOM: Y=RND
(7):R=100*(Y+10):GOTO R
1000 N=1:GOTO 2000
1100 N=2:GOTO 2000
1200 N=3:GOTO 2000
1300 N=4:GOTO 2000
1400 N=5:GOTO 2000
1500 N=6:GOTO 2000
1600 N=7:GOTO 2000
1700 N=8:GOTO 2000
2000 IF Y/3-INT (Y/3)=0THEN GOTO 2020
2010 LPRINT N; :GOTO 20
2020 LPRINT " f";: GOTO 20
```

The printed output will then consist of sequences containing letters "f" to represent the 'failures' to satisfy instruction 2000 (which rejects Y=N-1 multiple of 3), and the accepted values of N.

An example is: 5 6 5 f 6 2 5 2 f f 5 2 f 8 2 6 6 6 3 f f 6 5 f f 6 2 f f 6 2 2 3 5 6 f 5 5 f f 5 f f 5 8 5 8 3 3 2 3 2 6 2 3 3 f 2 f f 2 f 3 f 3 f f f f 6 3 2 f f 5 3 f 2 6 f 6 3 f 8 6 2 3 f 6 3 3 2 8 f f 2 2 f 2 3 6 5 6 8 3 2 3 3 6 f f 5 f 6 f 6 f 5 f 6 6 3 6 f 6 f f f f 3 5 6 5 3 6 f 3 2 2 3 5 f 2 f 6 2 8 f 6 f 3 f 8 6 8 5 5 6 2 5 6 2 8 8 2 8 8 3 f 8 8 f 8 2 5 8 8 6 f f 6 3 8 f 5 f 2 5 f 6 8 5 f 5 8 2 f 6 5 2 f 5 3 6 2 3 3 f, etc.

Of course, the above program can be made to simulate much better the interplay of random mutations and selection, but it is already sufficiently instructive as it stands, for it does simulate the essential features of the problem we are studying. The function RND(X) establishes the number of different states that will follow the 'present' state: if, instead of 7, we had had 0 and instruction 2000 had imposed a condition always satisfied (e.g. "N must be a real number"), then complete determination would have held. Total randomness (chaos) would have required X to be equal to the number of all admissible states, and, again, in instruction 2000 just the condition that the state produced should be an admissible physical state. As it stands, the program illustrates partial randomness (only one state out of eight), corrected by a sort of 'Darwinian selection'.

The analogy with a computer is perhaps somewhat unfamiliar to many philosophers of science. On the other hand, the mathematical and conceptual implications of the simulation of physical stochastic processes by mathematical models are a broad field of research, and our simple example must be considered as a quick way for hinting at the profound connection existing between the state-space conception of chance and causality and the rigorous formalisation and detailed analyses already available in general system theory (Casartelli, 1988; Schuster 1984).

6. A RE-INTERPRETATION OF CAUSE AND CHANCE IN THE PHYSICAL SCIENCES.

We have thus reached a point where it should be clear that, as far as the natural sciences are concerned, the concepts of cause and chance can be usefully re-interpreted in terms of the concepts of time flow, history, and compatibility, in full agreement with Lalande's remark quoted above. It appears that a scientific law (i.e. what is often called, in a broad sense, a 'physical' law) is essentially a statement about conservation or invariance of some quantity (e.g. energy), and provides a condition for the "compatibility" of the state B at time t' with the state A at time t. If the set of all applicable physical laws specifies uniquely the state B which will follow state A after a vanishingly small interval of time, we shall say that at time t we have a completely deterministic situation.

If complete determination holds for all the states realised during a finite interval of time starting at t and ending at t', then we can say that the state at time t' contains the whole "history" of the given physical system at least starting from time t. Of course, in this case such a statement expresses a trivial truth, because any state realised between time t and time t' contains the same information, and we can reconstruct the past and the future with equal certainty from the state at any instant between t and t'.

The same consideration ceases to be trivial in more general cases. Suppose that two states B and B' are allowed by the laws governing matter at time t' for a system initially in state A; we are in a case of partial determination. Suppose further that also at a time t" determination is not complete, and moreover the states C and C' compatible with B are not the same as the states D and D' compatible with B'.

Suppose now that at time t" we find the system, say, in state D. Then we may have two different situations. If the same chain as held from A to D holds in the direction of the past (viz., D and D' can only originate from B', C and C' can only originate from B, etc.), then we shall know that at time t' the system was in state B' and at time t it was in state A: at t" the system 'remembers' as it were its history, and its state contains non trivial information about the past. This situation is radically different from complete determination in the direction of the future, and is fully deterministic in the direction of the past: we are dealing here with a case very important for the well known problem of the arrow of time (Whitrow 1980, Kroes 1982). Suppose on the contrary that, in the direction of the past, D is compatible not only with B', but also with other states E, F, G, ..., and that the same holds for D', C, C'. Then, we may say that the very realisation of state D has erased (wholly or in part) the 'memory' of the past.As concerns chance, the former possibility illustrates the point that chance is strictly connected with time and the direction of its flow. In addition, both possibilities illustrate the fact that chance is related to incompleteness of 'objective' information, in the following sense: even when back-determination applies, the very fact that a state different from the observed one was compatible with the same initial conditions implies that the observed state cannot contain all the possible (= scientifically relevant) information about the given system. This point - however strange it may seem that what might have been has a relevance to reality - is a key point in discussions on epistemological issues in quantum mechanics, such as the famous non- locality problem (d'Espagnat 1979). A simple example is the following: suppose you have a system in a non-stationary state of energy E, and suppose you make just one energy measurement on it; then, according to quantum mechanics, you will find one of the energy values associated with the possible stationary states of the given system, at random. If the same experiment were repeated a sufficiently large number of times (or at the same time on a sufficiently large number of identically prepared systems), the average of the measured energy values would come as close as desired to E, so that energy conservation would be respected in the mean. On the other hand, since just one measurement is in question -for we are considering a single physical system in its actuality- information about the average will be completely absent.

We can now turn to a fundamental question: why is chance an essential ingredient of scientific explanation? We have already quoted Born's reflection, which certainly provides part of the answer. The rest of the same answer, in our opinion, requires the concept of 'irrelevance' on which we based previous studies (Del Re 1984), and which is a key concept in Salmon's (1984) analysis of causality. Randomness must be assumed whenever there are in (our description of) reality details that are irrelevant for the application of physical laws holding at a given level of reality. In the above example, specification at the atomic scale of the system on which a given energy measurement has been made is an irrelevant detail as far as the law of energy conservation is concerned, for that law is cogent only for systems of a much larger scale, consisting of very large numbers of atomic systems. Since there is no equivalent law at the atomic scale, a single measurement will produce any state out of the set of all states compatible with the observed system and the measuring apparatus. In other words, irrelevance is the source of randomness (in the natural sciences) in the sense that, whenever chance is called into play, this is a sign that we (or the probing system, which is the same because we are not interested here in conscious perception) are "asking questions" which are pointless as far as events at the given level of reality are concerned.

How does irrelevance apply to Aristotle's example (once again treated as if it described a sequence of events in the purely material sense)? Irrelevance arises here from the fact that neither of the two causal chains takes into account the existence of the other physical system (represented by either the debtor or the creditor) and of the other causal chain, until the two 'universes' merge into one: only from that event on will events in one system acquire relevance for the other system. The example is also useful to emphasize once again the role of information in connection with chance: some observer having information on both systems will not view the encounter as due to chance: the question is whether the existence of such an observer is admissible within a scientific scheme.

There is here a strong hint that the whole problem of cause and chance may look different if we change the rules of the game i.e. we go out of science into the wider world of natural philosophy, and from it to metaphysics.

7. IN THE DOMAIN OF METAPHYSICS.

We conclude this note by a quick glance at that wider field of reflection, and consider briefly the question: Is chance acceptable at a metaphysical level?

Let us first of all quote the answer given by T.S. Eliot (1985):

Time past and time present / are both perhaps present in time future / and time future contained in time past. / If all time is eternally present / all time is unredeemable. / What might have been is an abstraction / remaining a perpetual possibility / only in a world of speculation.

The above lines point out a most important fact: from the viewpoint of ontology time cannot be divided into past and future, and the only distinction that really matters is that between being and not- being. The problem of comprehending (not just understanding) reality is no longer one of history or information, it is one of necessity or contingency. The point is not whether or not some reality could have been different, for the only alternative to "is" is "is not"; the point is whether a given reality exists by itself or not.

Nevertheless, since concepts expressed by the same word must have something in common, there must be some common point at least in the two meanings of the attribute 'necessary': which means either not contingent or not random. An entity is necessary if it does not depend on any condition for existing; an event is necessary if there is no condition about its realisation, except the realisation of the event which 'causes' it. Thus, the ambiguity arises from the subtle distinction some philosophers make between existing and 'being realised'. If we consider the extension in space-time as just a property of reality, and events just as accidents of something that exists, then contingency must be distinct from randomness: the latter is a feature of extension in time much as angular points are a feature of shape, viz. of extension in space; the former is a property of existence as such. If, on the other hand, we assume that matter, with time and space, are the ultimate reality, then of course we can use the attribute contingent to mean random. But then it might be better not to speak of contingency at all, or to put it in parentheses, as Bunge does in his book on causality.

University of Naples (Italy).

BIBLIOGRAPHY

Aristotle (1961), Physique. Paris: Belles Lettres.

Born, Max (1951), Natural Philosophy of Cause and Chance. Oxford: Clarendon Press.

Bunge, Mario (1987), Kausalitaet: Geschichte und Probleme. Tuebingen: Mohr. Revised German ed. (transl. H. Spengler): Causality: the place of the causal principle in modern science. New York: Dover Publ. 1979.

Casartelli, M. (1988), The contribution of A.N. Kolmogorov to the notion of entropy. This book.

Del Re, G. (1984), Frequency and Probability in the Natural Sciences. Epistemologia 7 (spec. issue):75.

Eliot, T.S. (1985), Burnt Norton. In: e complete poems and plays. London: Faber and Faber 1985.

Espagnat, B. d' (1979), A la recherche du réel: le regard d'un physicien. Paris: Gauthier-Villars.

Kant I. (1956), Kritik der reinen Vernunft (1781). Felix Meiner Verlag, Hamburg.

Kroes, P. (1982). An inquiry into the structure of physical time. Nijmegen: Krips Repro Meppel.

Lalande A. (1968), Vocabulaire technique et critique de la philosophie. Paris: PUF.

Salmon, C. Wesley (1984), Scientific explanation and the causal structure of the world. Princeton, N.J.: Princeton University Press.

Schuster, P. (ed.) (1984), Stochastic phenomena and chaotic behaviour in complex systems. Berlin-Heidelberg: Springer.

Whitrow, G.J. (1980), The natural philosophy of time. Oxford: Clarendon Press.

John Archibald Wheeler

WORLD AS SYSTEM SELF-SYNTHESIZED BY QUANTUM NETWORKING[1]

The quantum, strangest feature of this strange universe, cracks the armor that conceals the secret of existence. In contrast to the view that the universe is a machine governed by some magic equation, we explore here the view that the world is a self-synthesizing system of existences, built on observer-participancy via a network of elementary quantum phenomena. The elementary quantum phenomenon in the sense of Bohr, the elementary act of observer-participancy, develops definiteness out of indeterminism, secures a communicable reply in response to a well-defined question. The rate of carrying out such yes-no determinations, and their accumulated number, are both minuscule today when compared to the rate and number to be anticipated in the billions of years yet to come. The coming explosion of life opens the door, however, to an all-encompassing role for observer-participancy: to build, in time to come, no minor part of what we call *its* past - *our* past, present, and future -but this whole vast world.

1. THE WORLD: A GREAT MACHINE OR A GREAT IDEA?

What is the structure of the world? Machinery, in the shape of a magic equation governing a geometry-like field in a supersymmetric manifold of ten or some other magic number of dimensions? Or an idea so obvious that is not obvious?

Idea? To illustrate the flavor of that word, one idea plus one model of that idea - inadequate, incomplete, and conceivably totally incorrect - is worth a hundred generalities. Let one such idea-plus-model serve as background for all that follows. Its information-theoretic character, the perspectives it suggests, the

E. Agazzi (ed.), Probability in the Sciences, 103–129.

issues it raises, and the probing questions that Rolf Landauer asks about it provide occasion to report it here. The idea? The world is a self-synthesizing system of existences. The model of how such a self-synthesizing system might be conceived to operate? The meaning circuit of **Figure 1**. That system of shared experiences which we call the world is viewed as building itself out of elementary quantum phenomena, elementary acts of observer-participancy. In other words, the questions that the participants put - and the answers that they get - by their observing devices, plus their communications of their findings, take part in creating the impressions which we call the system: that whole great system which to a superficial look is time and space, particles and fields. That system in turn gives birth to the observer-participants.

An "idea account" of the world of intercommunicating existences, one based on quantum-plus-information theory: How should it be viewed as relating to a continuum-plus-field-theoretic analysis? Not contradictory, but mutually illuminating. We do not say, "Thermodynamics is wrong, statistical mechanics is right". To do so, to deny the mutually supportive relation between these two outlooks on heat science, would be a total misunderstanding. Similarly here, between two very different views of the world - magic equation and magic idea - the future must be expected to bring us, not contradiction, but mutual illumination.

It is strange business to report about what we don't know. It is no stranger, however, than recounting the first half of a detective story of which the second half is missing. We know how difficult it is to pick out the clues, let alone assess them, unless we marshall them against the background of an idea. The idea here? Existences form a self-synthesizing system. The clues? Four stand out. Let us first list them, with brief commentaries. Then let us go back over each clue more carefully, asking how it bears on the suspicion that the quantum is the foundation of physics, that the world is a self-synthesizing system.

1. *No continuum*. Modern mathematical logic denies the existence of the conventional number continuum. Physics can do no other but follow suit. No natural way offers itself to do so

Figure 1

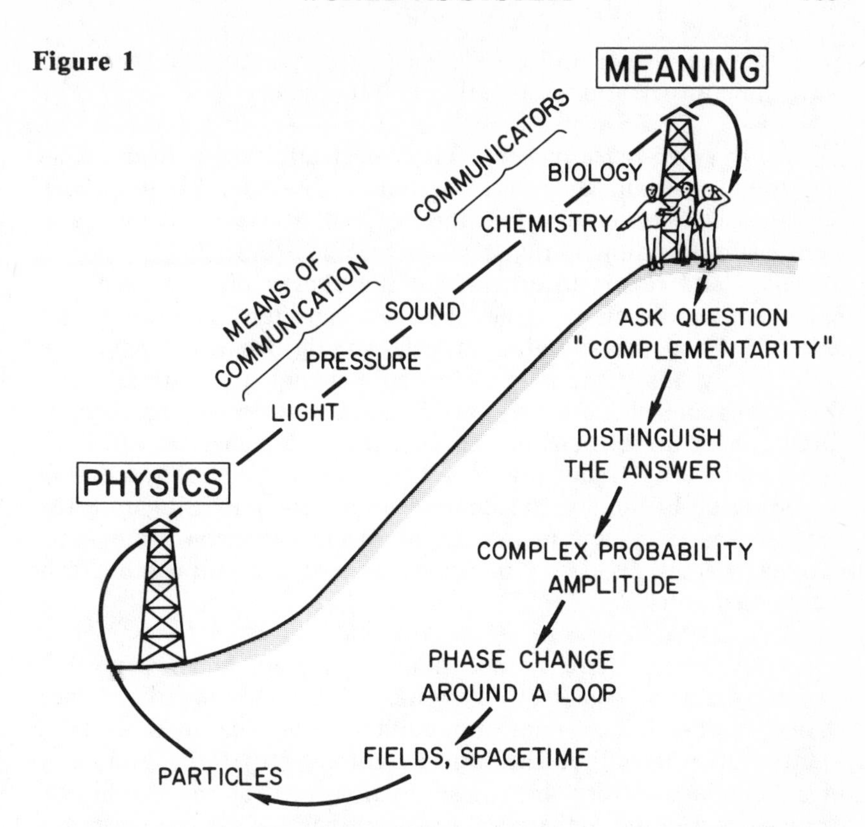

World viewed as a self-synthetizing system of existences. Physics gives light and sound and pressure - tools to query and to communicate. Physics also gives chemistry and biology and, through them, observer-participators. They, by way of the devices they employ, the questions they ask, and the registrations that they communicate, put into action quantum-mechanical probability amplitudes and thus develop all they know or ever can know about the world.
In a double-slit electron-interference experiment of the type proposed by Aharonov and Bohm, the interference fringes experience a phase shift proportional - so it is customary to say - to the flux of magnetic field through the domain bounded by the two electron paths. We reverse the language when we turn to the idea interpretation of nature. We speak of the magnetic field - and, by extension, spacetime and all other fields, and the whole world of particles built upon these fields - as having no function, no significance, no existence, except insofar as they affect wave phase, affect a 2-slit interference pattern, or more concretely, affect the counting rate of elementary quantum phenomena. Fields and particles give physics and close the loop.

except to base everything on elementary quantum phenomena, with their information-theoretic yes-no character.

2. *Observer-participancy.* The electron's momentum, the electron's position, do not exist out there independent of us. Not until we have installed and set the observing device and found what reading it registers do we have the right to say to ourselves and report to others that the chosen physical quantity had such and such a value. This is the inescapable sense in which we are participators in establishing what we have the right to say about the past. Minuscule though the part is today that such acts of observer-participancy play in the scheme of things, there are billions of years yet to come. There are billions upon billions of living places yet to be inhabited. The coming explosion of life opens the door to an all-encompassing role for observer-participancy: to build, in time to come, no minor part of what we call its past - our past, present and future- but this whole vast world.

3. *Austerity*: There is not one great field theory, neither electrodynamics, geometrodynamics, chromodynamics, nor string theory, which does not capitalize on the mathematical identity, the triviality, the logical tautology that the boundary of a boundary is zero. In this sense almost all of the machinery of physics is built on almost no machinery. This circumstance invites us to believe that all of physics is built on no machinery at all; that existence operates on the principle of total austerity.

4. *Timelessness.* The deepest insights we have on time today come out of Einstein's 1915 and still standard theory of general relativity in its quantum version. This quantum geometrodynamics tells us that the very concepts of spacetime and of before and after break down at ultrasmall distances. In tomorrow's deeper dispensation, we know that time cannot be an entity primordial and precise supplied - as elasticity once seemed to be - free of charge from outside physics. Like elasticity, the very concept of time must be secondary, approximate, derived: derived from profound consideration of a quantum flavor.

As we take a closer look at these four clues, we ask: To what extent do they comport with the concept of the totality of all existences as a self-synthesizing system? To what extent do these four items of evidence create difficulties for this closed-circuit view of nature? What are some of the problems calling for further investigation?

We cannot rightfully proceed with this assessment of the "idea theory" of the world without at least one word about the beautiful modern developments on the other side of the divide, in the heartland of the machinery view of nature, the domain of grand unified field theory and string theory. There, at least, one measure of progress is available. The kind of mathematics to be called on is clear: that synthesis of algebraic and differential geometry given us by Atiya, Singer, and other leaders in the field. Of the findings available out of that mathematics, has physics put to use at most a tenth? Then that number, in default of any other, tells something of our headway. On the idea side of the divide, however, we do not even know what the mathematics is, except that it cannot but be based on the integers and capitalize, surely, on information theory and on the guiding principle of many-body physics, "More is different".

2. NO CONTINUUM

The continuum of number theory: Who could dispense with it who works with matter and motion, particles and fields, space and time? Yet Hermann Weyl, who in earlier years took the concept of the continuum so seriously that he published a great treatise on the subject, in later years reversed his position, explaining, "(L.E.J.) Brouwer made it clear, as I think beyond any doubt, that there is no evidence supporting the belief in the existential character of the totality of all natural numbers." More generally he adds, "Belief in this transcendental world (of mathematical ideas, of propositions of infinite length, and of a continuum of numbers) taxes the strength of our faith hardly less than the doctrines of the early Fathers of the Church or of the scholastic philosophers of the Middle Ages".

Kurt Gödel, commonly identified as an idealist in mathematical logic in contrast to the constructivist Brouwer, nevertheless reported to his biographer Hao Wang, regarding the construction of the mathematical line, "According to this intuitive concept, summing up all the points, we still do not get the line; rather the points form some kind of scaffold on the line".

Willard Van Orman Quine, speaking from the world of mathematical logic, goes further: "Just as the introduction of the irrational numbers... is a convenient myth [which] simplifies the laws of arithmetic... so physical objects are postulated entities which round out and simplify our account of the flux of existence... The conceptual scheme of physical objects is [likewise] a convenient myth, simpler than the literal truth and yet containing that literal truth as a scattered part."

In brief, the physical continuum, and with it all the beautiful machinery of physics, is myth, is idealization. Existence, what we call reality, is built on the discrete.

Puzzle number one: If the world is founded on the discrete, why does every workaday description of it have to employ the continuum?

3. THE LESSON OF THE ELEMENTARY QUANTUM PHENOMENON

To the discreteness lesson of Weyl and Quine, out of the worlds of mathematics and logic, nothing in all of physics says a more vigorous "yes" than the elementary quantum phenomenon. There is not a sight we see, a pressure we feel, a sensation we detect which does not go back to elementary quantum phenomena for its explanation. On the discrete yesses and nos of these elementary quantum phenomena, on these iron posts of observation, we plaster in the papier-mâché of the continuum by an elaborate work of imagination and theory. However, despite this apparent continuum of every experience-the quantum teaches us- the world has at bottom an information-theoretic character.

No piece of the puzzle lies closer to hand than the quantum. In a letter of 1908 to his friend Laub, Einstein was already urging, "This quantum business is so incredibly difficult and important that everyone should busy himself with it." But how come the quantum? Out of what deeper idea derives its necessity in the construction of existence?

The quantum character of nature it is natural to assess differently according to whether one adopts the machinery or the idea vision of nature. In the machinery view, it is the role of the quantum to supply a rule for quantizing the master equation. In the idea view, the quantum cracks the armor that hides the secret of existence.

For a new understanding of how information fits into the scheme of things, we are indebted to no one more than Rolf Landauer. His work, and that of Szilard, Christodoulou, Bekenstein, Hawking, Fredkin, Toffoli, Bennett and others, has created new ties among information as bits, information as negentropy, information as mass-energy, and information as elementary quantum phenomena. The way of thought of information theory, we nevertheless can believe, will be of as much help in the new enterprise - to understand self-synthesis as plan without plan - as it already has been in the task of explaining entropy in terms of the elementary yes, no actions of the famous demon. More than one distinguished investigator - Kelvin, Maxwell, Szilard, Landauer, and Bennett - had to contribute an important idea before the final point became totally clear: The "thermodynamically costly act, which prevents the demon from breaking the second law, is not (as is often supposed) the measurement by which the demon acquires information about the molecule being sorted, but rather the resetting operation by which this information is destroyed in preparation for making the next measurement." We are seeing the dawn of a new third era of physics:

Era I - Motion with no explanation of motion: the parabola of Galileo and the ellipse of Kepler.

Era II - Law with no explanation of law: Newton's laws of motion, Maxwell's electrodynamics, Einstein's geometrodynamics, modern chromodynamics, grand unified field theory, and string theory.

Era III - Information-based physics.

No feature of quantum theory is more central than the complex probability amplitude, no question more frequently asked than "How come this complex probability amplitude?" and no answer more satisfying than that given by information theory at the hands of R.A. Fisher, E.C.G. Stueckelberg, and W.K. Wootters. That answer has two parts: the asking of a question and the distinguishing of an answer.

Fisher found himself forced into a probability amplitude - a real probability amplitude - by his premodern-quantum-theory 1922 work in the field of population genetics. This work Wootters clarified, extended, and generalized in 1980 Ph.D. thesis, in which he also spells out the relation to quantum theory.

An example? We find ourselves in the midst of a tribe of people who speak an unknown language. Are they the Eddas, who are friendly? Or are they the Thors, who are cannibals? All we have to go on is the color of the eyes of the sixteen warriors who encircle us. Our scouts have told us that 67.3 percent of the Eddas have grey eyes; 32.7 percent, blue eyes; whereas for the Thors the proportions are the other way around. Our statisticians have told us that, if the majority of the sixteen pairs of eyes are grey, we have close to a twelve-to-one chance of being safe. And so they are- and so we are! That is distinguishability in action.

Unfortunate explorers, we find ourselves on a new journey to a new continent confronted anew by the old issue. Are the sixteen who now surround us the friendly Aeolians or the deadly dangerous Boreans? At first sight, it appears that it will be much more difficult to be certain of our appraisal. Why? Because the differences are now so much less between the two tribes in count of grey and blue yes.This conclusion bases itself (plane $p_{grey} + p_{blue} + p_{brown} = 1$ in the upper left-hand diagram of Figure 2) upon the separation of the two representative points in question in a linear probability diagram, a separation large in the one continent, small in the other.

Statistical analysis, however, shows that if the grey eyes are again in the majority, we again have close to a twelve-to-one assurance of being safe. The linear diagram is misleading because it is based on probabilities. To make distinguishability properly shine out, we should use not probabilities but probability amplitudes; not linearly related quantities that lie on a sector of a plane, but quadratically related quantities that lie on a sector of a sphere.

$$(p_{grey}{}^{1/2})^2 + (p_{blue}{}^{1/2})^2 + (p_{brown}{}^{1/2})^2 = 1.$$

In brief, the proper depiction of distinguishability demands Hilbert space. The angle in Hilbert space between two nearly identical probability-amplitude vectors

Figure 2

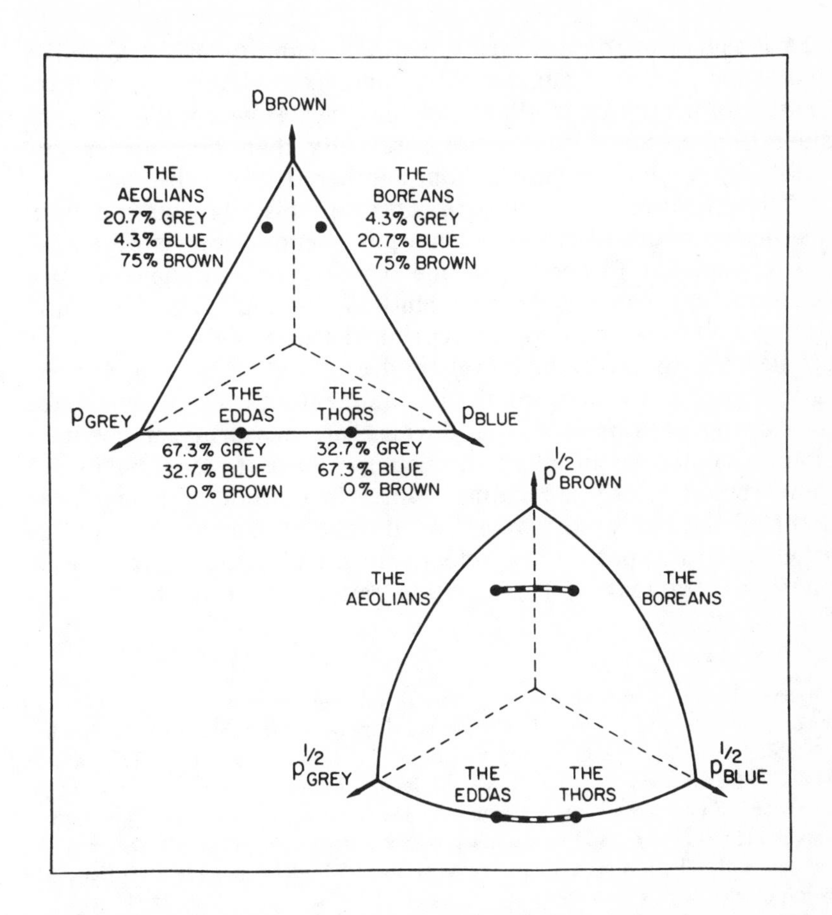

From probabilities to probability amplitudes as tool for determining distinguishability. Triangle above: probabilities of gray, blue, and brown eyes for tribes plotted in three-dimensional probability space. Quarter-sphere below, Hilbert space: same information with axes now measuring "probability amplitudes". The angle (dashed arcs) between two points in this Hilbert space measures the distinguishability of the two populations. W.K. Wootters is thanked for assistance in preparing this diagram.

(stippled lines in lower right-hand diagram, Figure 2), Wootters shows, is the proper measure of their distinguishability.

The Fisher tool for measuring distinguishability, his probability amplitude, is a real number. The complex probability amplitude of quantum physics is a complex number. How come? No consideration presents itself more forcefully than this: Fisher, distinguishing one population from another, is pure observer.

The quantum-level experimenter, or his observing device, dealing with the elementary quantum phenomenon, is observer-*participant*. For Fisher to ask one typical question about his population, eye color, does not stand in a complementary relationship to his asking another typical question, height. To ask of an electron its position, however, does stand in a complementary relation to demanding its momentum. The device for measuring position and the device for measuring momentum simply cannot be installed in such a way as to operate in the same region at the same time. More: In default of a measurement of the one or the other, we do not even have the right to attribute either position or velocity to the electron. No choice of question? No answer! Participation reveals itself in the demand for choice.

Observer-participancy, showing as it does in this requirement for choice, and belonging as it does to the world of the small displays some analogy to the familiar game of "find the word in twenty questions" in its surprise version. The ones to whom I must put my questions have - unbeknownst to me - agreed not to agree on a word. Each answers my question by a yes or no as he pleases - with one small proviso. If I challenge, and he cannot produce a word compatible with his own reply and with all previous answers, he loses and I win. The game is as difficult for everyone else as it is for me.

Is the word sitting there in my friends' custody, waiting for me, as I suppose, when I enter the room? No. The word with which we end up is not even on the docket before I choose and pose my questions. In this game, as in quantum physics, no question, no answer. What word comes out, moreover, depends on my choice of questions. Different questions? Or the same questions in a different order? Different outcome. The outcome, however, does not depend on my choices alone. My friends also have a hand in it, through their selection of answers. In summary, the game of twenty questions, in its surprise version, promotes me from observer to observer-participant.

Observer-participancy is the central feature of the world of the quantum. We used to think of the electron in the atom as having a position and a momentum whether we observed it or not, as I thought the word already existed in the room whether I guessed it or not. But the word did not exist in the room ahead of time, and the electron in the atom does not have a position or a momentum until an experiment is conducted to determine one or the other quantity. The questions I asked had an irretrievable part in bringing about the word that I found - but I did not have the whole voice. The determination of the word lay in part with my friends. They played the role that nature does in the typical experiment, where so often the outcome is uncertain, whether with electron or with photon. In brief, complementarity symbolizes the necessity to choose a question before we can expect an answer:

> *Complementarity*: "...any given application of classical concepts precludes the simultaneous use of other classical concepts which in a different connection are equally necessary for the elucidation of the phenomena." (Bohr's abbreviated 1934 version of the principle of complementarity, propounded by him in his famous fall 1927 Como lecture to penetrate what Heisenberg had not penetrated in his spring 1927 principle of indeterminacy).

We once thought, with Einstein, that nature exists "out there", independent of us. Then we discovered - thanks to Bohr and Heisenberg - that it does not.

Not all of the surprises hidden in the quantum had to be uncovered to reveal how it comes about that the probability amplitude of the quantum world must be a complex number. This discovery Stueckelberg published in 1960. He used as foundation for his argument Heisenberg's spring 1927 principle of indeterminacy. The key point in the reasoning, however, we realize in retrospect, was complementarity, complementarity in the sense of choice: No choice, no answer. Complementarity stands revealed as the cryptic message of Schrödinger's complex-valued ψ.

Complementarity was not the last idea feature of nature to be revealed in the quantum. Bohr had to enunciate a further concept in 1935 to cope with the issue about "reality" raised by Einstein, Podolsky, and Rosen earlier that year. This is the elementary quantum phenomenon, "brought to a close" by an "irreversible act of amplification".

It would be difficult to give an example of an elementary quantum phenomenon simpler than the split-beam experiment of Figure 3. Twenty-four photons enter in a twenty-four-hour day. Before we analyze what happens, let us describe it in wrong but at first sight tempting language: Half of the photons, on the average, penetrate the first half-silvered mirror. They follow the low road to the lower total reflector. They bounce up to trigger the detector at the upper right. The other photons, twelve on the average, are reflected at the first half-silvered mirror. They follow the high road and set off the counter at the lower right. Insert, however, the second half-silvered mirror. Give it a well-chosen elevation. Then we ensure mutual cancellation of the two partial waves on their way to the counter at the upper right, one of them the reflected wave that has come from the high road, the other the transmitted wave that has come from the low road. That counter registers not at all. In contrast, the two partial waves traveling to the counter at the lower right have identical phase. They totally reinforce. All twenty-four photons arrive at the lower counter.

Treat the same photons sometimes as waves and sometimes as particles? Surely quantum mechanics is logically inconsistent! This was Einstein's position in the first phase (1927-1933) of his twenty-eight-year-long wrestle with modern quantum theory. Schrödinger, too, expressed his unhappiness, saying that if he had known of all this Herumspringerei - all this jumping about between wave and particle interpretations - to which quantum theory would lead, he would never have had anything to do with it in the first place. Bohr's reply to both was simple. We can leave out the second half-silvered mirror, or we can put it in. However, we can't do both at the same time. Complementarity, yes; contradiction, no.

Querying Bohr one evening, his favorite professor, the old Danish philosopher Harald Hoffding, put to him this question about a similar and even better known idealized test case, the double-slit experiment: "Where can the photon be said to be in its passage from the slit to the photographic plate?" "To be?" Bohr replied, "To be? What does it mean, 'to be'?"

The same question poses itself with even greater force in the delayed-choice version of either the double-slit or the split-beam experiment. We can delay our choice whether to put in the second half-silvered mirror or to leave it out. We can delay that choice until the photon has passed through the first half-silvered mirror, has undergone reflection at the next mirror, and has arrived almost at the point of crossing of the two beams. To interpose this delay in our choice makes no difference in the outcome.

This, theory tells us; and this, independent delayed-choice experiments in three different laboratories already confirm. This finding shows how wrong it is to say that, with mirror out, we find out "which route" the photon traveled, or, with mirror in, what the difference in phase is in a "two-route mode of travel". It is wrong to speak of what the photon is doing between the point of entry and the point of registration.
The right word, Bohr emphasized is, phenomenon. In today's words, no elementary quantum phenomenon is a phenomenon until it is a registered phenomenon - that is, indelibly recorded or brought to a close, in Bohr's phrase, by an irreversible act of amplification, such as the avalanche of electrons in a Geiger counter or the blackening of a grain of photographic emulsion or the click of a photodetector.

The elementary quantum phenomenon is a great smoky dragon. The mouth of the dragon is sharp where it bites the counter. The tail of the dragon is sharp where the photon enters. But about what the dragon does or looks like in between we have no right to speak, either in this or in any delayed-choice experiment. We get a counter reading, but we neither know nor have the right to say by what route it came.

Normally the quantum dragon operates so far beneath the everyday hardware of physics that we have to pursue it to its lair to catch it biting. Thus, this plank we see yielding a little as we sit on it. This yielding we interpret as elasticity. This elasticity we understand in terms of the linkage between molecule and molecule intermediated by hydrogen bonds. A single hydrogen atom we can arrange to detect in the laboratory. When finally we speak of the electron of a single atom as residing in this, that, or the other quantum state of excitation, we begin to close in on the dragon. No bite yet, however; still totally smoky. As smoky as the photon is in the split-beam experiment before we have put photodetectors in the way to register its arrival, so smoky is the dragon that we call the electron before we have arranged equipment to get it out where it can accomplish an irreversible act of registration - or to get out and register some equivalent entity, such as a photon, that has interacted with it. Only with this registration do we have the basis for the elementary question plus the yes-or-no answer of observer-participancy.

The elementary quantum phenomenon is the strangest thing in this strange world. It is strange because it has no localization in space or time. It is strange because it has a pure yes-no cha

Figure 3

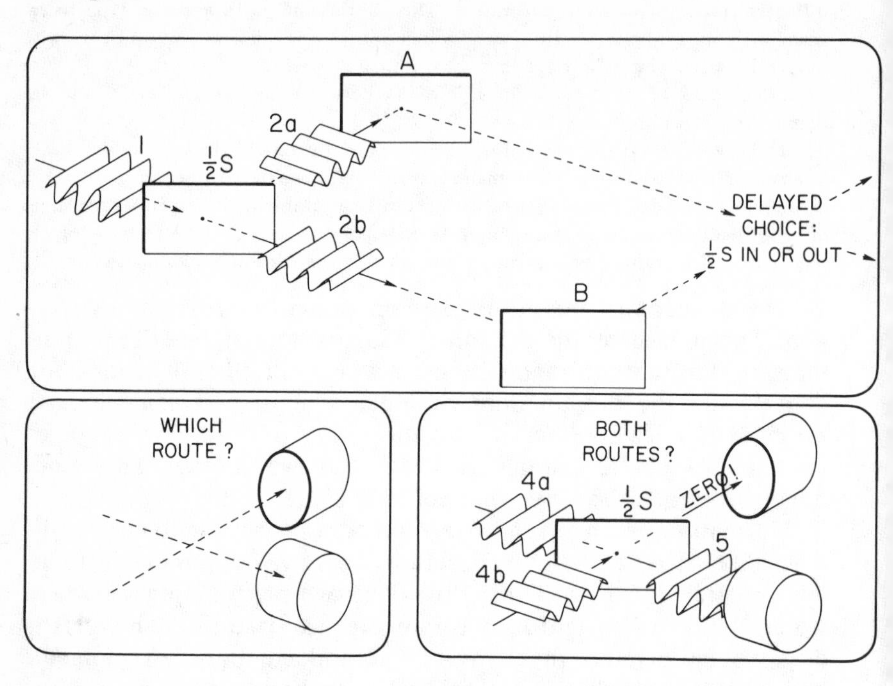

Beam splitter (above) and its use in a delayed-choice experiment (below). In the arrangement at the lower left, half of the photons, on the average, go into the upper counter and are registered there. However, when a half-silvered mirror is introduced and properly positioned (lower right), that counter gives zero counts. All the photons go to the counter at the lower right. The choice whether to put in the half-silvered mirror or to take it out can be made at the very last minute. It is wrong to say that one decides, after the photon has "already done its travel," that the photon "has come by one route" (or "by both routes"). The photon is a great smoky dragon, its teeth sharp where it bites the one counter or the other, its tail sharp at its birthplace, but in between totally smoky.

racter- one bit of meaning. It is strange because it is more deeply dyed with an information-theoretic color than anything in all physics. This strangeness makes it natural to ask not only what lies behind and beneath the elementary quantum phenomenon, but also - *puzzle number* two - what role it has in building all that is.

How subtle the divide is between what we call somethingness and nothingness! That lesson of the elementary quantum phenomenon we see in a new form when we turn to the role played in the construction of field theory by the principle that the boundary of a boundary is zero.

4. AUSTERITY

"So far as we can see today, the laws of physics cannot have existed from everlasting to everlasting. They must have come into being at the big bang. There were no gears and pinions, no Swiss watchmakers to put things together, not even a preexisting plan. If this assessment is correct, every law of physics must be at bottom like the second law of thermodynamics, higgledy-piggledy in character, based on blind chance.

"There is no simpler illustration of the second law than the way molecules distribute themselves between two regions in proportion to the volumes of those two regions.... Every heat engineer knows he can design his heat engine reliably and accurately on the foundation of the second law. Run alongside one of the molecules, however, and ask it what it thinks of the second law. It will laugh at us. It never heard of the second law. It does what it wants. All the same, a collection of billions upon billions of such molecules obeys the second law with all the accuracy one could want. Is it possible that every law of physics, pushed to the extreme, will be found to have the character of the second law of thermodynamics, be statistical and approximate, not mathematically perfect and precise? Is physics in the end 'law without law', the very epitome of austerity?

"Nothing seems at first sight more violently to conflict with austerity than all the beautiful structure of the three great field theories of our age, electrodynamics, geometrodynamics, and chromodynamics (under which latter heading, for convenience's sake, we speak also of Klein-Kaluza and string theory in their various forms). They are the fruit of hundreds of experiments, scores of gifted investigators and a century of labor. Impressive treatises spell out the physics and mathematics of all three theories. How can anyone possibly imagine all this richness coming out of a higgledy-piggledy origin?

"Only a principle of organization which is no organization at all would seem to offer itself. In all of mathematics, nothing of this kind [is] more obviously [available] than the principle that 'the boundary of a boundary is zero' [or, in mathematical terminology, $\partial\partial = 0$]. Moreover, all three great field theories of physics use this principle twice over, once in the form that 'the one-dimensional boundary of the two-dimensional boundary of a three-dimensional region is zero', and again in the form that 'the two-dimensional boundary of the three-dimensional boundary of a four-dimensional region is zero' (or the pair of equivalent higher-dimensional statements in any version of field theory that operates in a higher-dimensional manifold). This circumstance would seem to give us some reassurance that we are talking sense when we think of almost all of physics being founded on almost nothing." [J.A.W., *Physics and Austerity: Law Without Law*, Anhui Science and Technology Publications, Hefei, Anhui, People's Republic of China, 1982].

To spell out the boundary principle in the context of electrodynamics would be too trivial to serve as good illustration; in the domain of chromodynamics and string theory, too technical; but just right in geometrodynamics. Gravitation is a theory of intermediate difficulty, great interest, and widely recognized beauty. In it the central idea lends itself to statement in the single word "grip". Spacetime grips mass, telling it how to move. Mass grips spacetime, telling it how to curve.
What help do we get in understanding the grip of gravity from the principle that the boundary of a boundary is zero?
In older times we looked on mass or charge as primary, as source, as the ultimate entity, and regarded the gravitational or electromagnetic field as secondary. The source "knew" that it wanted to be conserved. The field ran along behind as slave, obedient to its wish. Today we regard the field as primary and the source as secondary. Without the field to govern it, the source would not know what to do. It would not even exist.

When two gigantic spaceships smash into each other, much is destroyed. One quantity, we know, is conserved, the energy-momentum 4-vector. What master is so powerful that it can hold those two mighty spaceships in straight-line motion before they hit and to see to the conservation law in the crash itself? Spacetime! Spacetime grips them both. Spacetime, right where they are, enforces the conservation of momentum and energy.

How does nature wire up momentum-energy - momenergy - the source, to spacetime geometry, the field, so as to guarantee conservation of the source, and do this automatically, without benefit of a corps of Swiss watchmakers, with no gears or pinions at all? By applying the principle $\partial\partial = 0$ at the 2-3-4-dimensional level.

It is automatic that there shall be no creation of source in the region of space $\Delta x \Delta y \Delta z$ during the time Δt. How so?

It is the first part of this question to ask what we mean by "no creation" in the 4-dimensional cube $\Delta x \Delta y \Delta z \Delta t$; that is, to enquire how we test for no creation. A look at the eight 3-dimensional faces of that 4-dimensional cube is enough to disclose the test. Evaluate the amount of momentum-energy contained within one of those eight 3-cubes, say the cube with dimensions $\Delta x \Delta y \Delta$t, located at the distance $\Delta x/2$ to the "right" of the center of the 4-cube. Make the same evaluation for each of the other seven 3-cubes. Add up the results, with due regard to sign. Is the total zero? Then conservation is upheld in this sense: The amount of momenergy in $\Delta x \Delta y \Delta z$ at the end of the time interval Δt differs from the amount $\Delta x \Delta y \Delta z$ at the beginning of that time interval by exactly the amount transported in across the six faces of $\Delta x \Delta y \Delta z$ during the time Δt. There must be no discrepancy, no creation of momenergy out of the emptiness of space.

Now for the remaining part of our question: By what automatic means does geometrodynamics meet this test of zero creation? Answer: It identifies the content of momenergy inside each 3-cube as the sum - with due regard to sign - of contributions from the six 2-faces that bound that 3-cube.

The detail? With eight 3-cubes, and six 2-faces per 3-cube, the 4-cube of concern to us presents us with 8 x 6 = 48 faces. Each of these faces makes its own individual contribution to the momenergy inside one or another of the 3-cubes; makes its own contribution, consequently, to the bookkeeping balance which is to tell us that there has been no creation at all inside $\Delta x \Delta y \Delta z$. A zero balance, yes. But how? Now it comes. These faces butt up against each other in pairs. Not a single face is exposed to the outside. Moreover, each face makes a contribution equal in magnitude and opposite in sign - conventions about sign being what they are - to the contribution of its partner face. Zero total, yes; and most remarkable of all, zero automatically. Conservation from a tautology, from a stupidity, from the central identity of algebraic geometry, $\partial\partial = 0$, in the form which states that the 2-dimensional boundary (here: those 48 faces...) of the 3-dimensional boundary of a 4-dimensional region is automatically zero.

Machinery? Absent. Absent from the law of conservation of source in not only gravitation but also every other great field theory. A closer look, however, discloses a residue of machinery. It gives each theory its own characteristic form.

In gravitation the key device is spacetime curvature. It associates a rotation with circum-navigation of any chosen 2-face of a 3-cube. The six faces of the elementary 3-cube thus present us with six rotations. Add them? Use that sum over the faces of the 3-cube to define the content of momentum-energy within the 3-cube? That dream collapses. To ask for the sum of those six rotations is to ask for the result of circumnavigating, in turn, the six 2-faces of the 3-cube. In the necessary travel we traverse each edge of the cube twice, once in one direction, then again in the opposite direction. Total cancellation results whenever the cube is small enough so that we can neglect second-order-terms. That is the boundary principle in action, not in its previously used 2-3-4-form but now in its 1-2-3 form: The 1-dimensional boundary of the 2-dimensional boundary of a 3-dimensional region is automatically zero.

Elementary mechanics teaches us to expect an object to start rotating even when the vector sum of the forces acting upon that object is zero. What counts in producing rotation is not the forces themselves but their moments. Moments about what point? That does not matter, because the vector sum of the forces is zero.

Similarly in geometrodynamics. We expect momenergy within a 3-cube even though the sum of the rotations over the faces of that 3-cube is zero. What measures the content of momenergy is not the rotations themselves but their moments. Moments of rotation about what point'? That does not matter, because the sum of the rotations is zero. In this elementary idea - due to the insight of Elie Cartan - we have before us the whole way of action of Einstein's great theory of gravity: the grip of spacetime on mass, and the grip of mass on spacetime.

Relativistic gravitation theory today is an enormous subject, reaching from the structure of black holes to the deflection of light, and from gravitational waves to cosmology. To understand all this physics, simple geometric constructions suffice. Nowhere is this geometric simplicity of the subject more evident than in its central device, the grip that couples mass with spacetime geometry, the moment of rotation.

Problem: If in gravitation and the other great field theories we can derive so much from so little, why can't we go the rest of the way and obtain everything from nothing? What holds us back from a physics of total austerity? Two obstacles, above all: dimensionality and time.

About dimensionality there was no doubt in the days from Newton to Einstein. There was doubt only about which one or another of a dozen arguments supplied the authentic magic derivation for the magic number three. A very different idea has received much instructive investigation in our own day: Yes, there is a magic dimensionality, but no, it is not three. The extra dimensions are envisaged as curled up into closure in a space so small as not to be susceptible to investigation by any everyday means. Particle masses appear as organ-pipe resonances in this microspace. The new question, What is the right dimensionality, has replaced the old question, What is the right derivation for three.

A third assessment imposes itself on us here: Nature, examined penetratingly, has no dimensionality at all. Dimensionality implies exactly what mathematical logic denies, the continuum. The appearance of a continuum, however, is undeniable, and with it the appearance of a dimensionality. It is difficult to appraise this apparent continuity and this apparent dimensionality as other than cover-up, plastered over a more subtle structure that has neither continuity nor dimensionality.

Puzzle number three: How are we to reconcile the demand for structure with the demand for total austerity?

5. TIMELESSNESS

When we appraise dimensionality as cover-up, when we rank continuum as illusion, then we must also interpret time as myth.

The concept of time was not handed down from heaven. Neither was it supplied free of charge from outside for the benefit of physics. The very word is a human invention, and the problems that come with it are of human origin. The miracle is only this, that a notion with so little undergirding has managed to stretch, without snapping, to encompass so much. Einstein's 1915 geometrodynamics continues to serve as the generally agreed authority for all that time now means and measures.

Time today is in trouble: (1) Time ends - Einstein's theory tells us - in big bang and gravitational collapse. (2) Past and future are interlocked in a way contrary to the causal ordering presupposed by time, in this sense: According to how the observing equipment in the here and now is set one way or another, that choice has irretrievable consequences for what we have the right to say about the past, even the past billions of years ago, before there was any life. The past has no existence except as it is contained in the records, near and far, of the present. (3) Quantum theory denies all meaning to the concepts of "before" and "after" in the world of the very small, at distances of the order of the Planck length, $L = (\hbar G/c^3)^{1/2} = 1.6 \times 10^{-33}$ cm. Spacetime is the classical history of space geometry evolving deterministically in time. The very notion stands in utter contradiction to the long-known lessons of complementarity and indeterminism about the quantization of any classical field theory. A proper quantum account of the dynamics of geometry teaches us that - except in the above-Planck-length approximation - there is no such thing as spacetime.

Table 1 The concepts of *universe* and of *multiple-existence world* compared and contrasted

	Universe	*World of existences*
Machine	Yes	No
Time	Yes	No
Record of change	Conditions on a sequence of spacelike hypersurfaces	Yes or no records of a multitude of observer-participators
Mathematics	Continuous fields	Discrete yes or no
Dynamics	Via machinery	Via asking questions

It is not enough in dealing with these difficulties to quantize Einstein's geometric theory of gravity according to the pattern for quantizing any other standard field theory; not enough to write down the resulting often-discussed wave equation:

$$\{-\partial^2\psi/[\partial^{(3)}G]^2\} + {}^{(3)}R\psi = 0;$$

not enough - despite all the fascination and instructiveness of the work of Everet, De Witt , Hartle, and Hawking towards interpreting the result - to calculate in this way the probability amplitude $\psi[^{(3)}G]$ for this, that and the other 3-geometry. This whole line of analysis presupposes that there is such a thing as "the universe".

Even to utter that noun is to hear as if it were yesterday Lord Rutherford standing in the Cavendish Laboratory hallway thundering, "When a young man in my laboratory uses the word 'universe', I tell him it is time for him to leave". We try to avoid the very concept of universe in the present account because of all the ideological presuppositions (**Table 1**) latent in the word. World: a multiplicity of existences? Yes. Universe? No.

The minuet? How harmonious, how fascinating, how beautiful. Yet all the while we watch we know that there is no such thing as a minuet, no adherence with perfect precision to a pattern, only individuals of different shapes and sizes pursuing different plans of motion with different accuracies. Let this clearer view suggest the totally different idea of a multi-existence world that the concept of observer-participancy would offer in place of the assumption-laden word *universe*.

The world *timelessness*, in summary, stands for the thesis that at bottom there is not and cannot be any such thing as time; that we have to expect a deeper concept to take its place. Events, yes. A continuum of events, no.

Puzzle number four: How to derive time without presupposing time.

6. THE WORLD OF EXISTENCES AS A SYSTEM SELF-SYNTHESIZED BY QUANTUM NETWORKING

No time, no law, no machinery and no continuum: Four clues more pregnant with guidance it would be difficult to imagine. Immensely more difficult is this - how to employ these clues, how to unravel the secret of existence, how to get *numbers* and *predictions*. If we have no answers, we have at least one encouragement. It generally carries us at least eighty percent of the way towards the solving of a deep puzzle to ask enough nearly right questions!

We see how powerful our four clues are when we compare and contrast the schematic diagram of Figure 1 for the world as a system self-synthesized by quantum networking with two other self-synthesizing systems, the modern worldwide telecommunications system and life.

Beginning with a single telegraph line connecting a single sender and a single receiver and expanding to a global multimode network, telecommunications constitute today an industry ever more immense in its extent. However, that growth is no machine. It is an immensity of demands and responses. The telecommunications industry is not telecommunication. The telecommunications industry is telecommunication plus life.

Only so could telecommunications become what it is today, a self-synthesizing system.

That other self-organizing system, life itself, likewise shows a fantastic complexity of structure. However, its marvels go back for explanation, we know, to mutation plus natural selection. Life, like telecommunications, is in a continual state of evolution.

Both self-synthesizing systems show this immense difference from existence -that they submit to time, the *outside* metronome which drives them- whereas elementary quantum phenomena leap across time and, on the Figure 1 model of world as self-synthesizing system, *generate* time. There are other differences, among them the following.

No place to "start." A closed circuit.

There was a toehold for the telecommunications system to start its self-synthesis: the community of potential communicators plus the expanding power of physics to provide new means of communication. Life, too, had a pre-existing foundation on which to build itself - chemistry in the fullest sense of the word chemistry. But the world of existences: where and when and on what foundation can it possibly be imagined to build itself? Might not one just as well speak of making of "airy nothing a local habitation and a name"?

Whoever would sail the craft of reason through the sea of mystery to find a foundation for existences has to steer his way between twin rocks of destruction: postulate an inexplicable something on which to build? That would shatter a central principle of Western thought: every mystery can be unraveled. Or postulate under each level of structure another, and under that yet another, in a never-ending sequence? That would be equally disastrous. No way offers itself to navigate a course between these rocks of ruin except to believe that the world of existences synthesizes itself after the pattern of a closed circuit.

Life, mind, communication count for nothing in the scheme of existence? Everything!

An elementary quantum phenomenon put to use to establish meaning: There's the rub. How can we reconcile such a life-and-mind-centered notion with the traditional spirit of physics? Einstein speaks of the inspiration of his youth, "Out yonder there was this huge world, which exists independently of us human beings and which stands before us like a great, eternal riddle..." Marie Sklodowska Curie tells us, "Physics deals with things, not people." David Hume asks, "What peculiar privilege has this little agitation of the brain which we call thought, that we must thus make it the model of the whole universe?"

Are life and mind indeed unimportant in the workings of existence? Is life never to inherit the vastness of space because today its dominion is so small? Or is not rather life destined to take possession of all the out-there because the time available for conquest is so large? How easy it is to be overimpressed by the remoteness of the quasars; how tempting to discount as anthropocentric any purported place for life and mind in the construction of the world. Is it not even more anthropocentric to take man's migration by foot and ferry in fifty thousand years as the gauge of where life will get in fifty billion years?

The fight against here-centeredness began with the 1543 *De revolutionibus orbium coelestium* of Copernicus. The time-bridging power of the elementary quantum phenomenon warns us today to battle against now-centeredness.

Life and mind: for how much can they be conceived to count in the scheme of existence? Nothing, say the billions of light years of space that lie around us. Everything, say the billions of years of time that lie ahead of us.

It cannot matter that man in time to come will have been supplanted by, or will have evolved into, intelligent life of quite other forms. What counts - in the idea view being explored in this paper - is the rate of asking questions and obtaining answers by elementary quantum phenomena, acts of observers-participancy, exchanges of information. If space is closed, if - following on the present phase of expansion - the system of galaxies contracts, if temperatures rise, all in line with the best known Friedmann cosmology, and if life wins all, then the number of bits of information being exchanged per second can be expected to rise enormously compared to that number rate today. The total count of bits: how great will it be before the counting has to cease because space is within a Planck time of total crunch? And how great must that future total be - tally as it is of times past - to furnish enough iron posts of observation to bear the smooth plaster which we of today call existence?

Bits needed. Bits available. Calculate each. Compare. This double undertaking, if and when it becomes feasible, will mark the passage from clues about existence to testable theory of existence.

No ensemble, no factory for making universes. Observer-participancy the whole source of the "out there" plus *life, mind, communication*

Counting bits is one test of theory for the future; accounting for the reciprocal fine-structure constant, $\hbar c/e^2 = 137.036...$, and the famous large-number dimensionless constants of physics is another. Those constants must have nearly the values they do, Robert H. Dicke, Brandon Carter, and others point out, if life is ever to be possible -not merely life as we know it, but life of almost any conceivable form. This observation has led some investigators to the idea of an ensemble of universes, one differing from another in the values of the dimensionless constants - a latter-day version of those words of David Hume from two centuries ago: "Many worlds might have been botched and bungled, throughout an eternity, ere this system was struck out: much labor lost: many fruitless trials made, and a slow, but continued improvement carried on during infinite ages in the art of world-making". There operates on such an ensemble of universes, Charles Pantin argued in 1951, something "analogous to the principle of Natural Selection, that only in certain Universes, which happen to include ours, are the conditions suitable for the existence of life, and unless that condition is fulfilled there will be no observers to note the fact". This ensemble concept is common to many of today's versions of the cosmological anthropic principle, reviewed in the comprehensive book of John D. Barrow and Frank J. Tipler (*The Anthropic Cosmological Principle*, Clarendon Press, Oxford 1986).

The contrast between the two views could hardly be greater: selection-from-an-ensemble and observer-participancy. The one not only adopts the concept of universe, and this universe as machine, it also has to postulate, explicitly or implicitly, a supermachine, a scheme, a device, a miracle, which will turn out universes in infinite variety and infinite number. The other takes as foundation notion a higgledy-piggledy multitude of existences, each characterized, directly or indirectly, by the soliciting and receiving of answers to yes-no questions, and linked by exchange of information.

Solipsism, no; communication, yes

Solipsism? Solipsism in the dictionary sense of "the theory or view that the self is the only reality"? Not so! We can even question whether two often-quoted thinkers of the past ever meant anything at all like solipsism in this sense by their well-known statements: Parmenides declaring that "What is ... is identical with the thought that recognizes it," and George Berkeley teaching that "*Esse est percipi*": to be is to be perceived. The heart of the matter is the word *self*. What is to be understood by the word self we are perhaps beginning to understand today as well as some of the ancients did. We know that in the last analysis there is no such thing as self. There is not a word we speak, a concept, we use, a thought we think which does not arise, directly or indirectly, from our membership in the larger community. On that community the mind is as dependent as is the computer. A computer with no programming is no computer. A mind with no programming is no mind. Impressive as is the greatest computer program that man has ever written and run, that program is as nothing compared to the programming by parents and community that makes a mind a mind.

The heart of mind is programming, and the heart of programming is communication. In no respect does the observer-participancy view of the world separate itself more sharply from universe-as-machine than in its emphasis on information transfer.

7. THE GREAT QUESTION

Will we ever succeed in stripping off the continuum, in comprehending the why of the quantum, in achieving a physics of total Einstein^austerity, in deriving -without time- the essence of time? And all this by interpreting the world as a self-synthesizing system of existences built on observer-participancy? In assessing this enterprise, we have the advice of Niels Bohr that "...every analysis of the conditions of human knowledge must rest on considerations of the character and scope of our means of communication."

Department of Physics, The University of Texas.

NOTES

This paper is dedicated to Rolf Landauer on the occasion of his 60th birthday and in special appreciation of an important lecture given by him, "Computation: A Fundamental Physical View" at the IBM United Kingdom Symposium, Science and Paradox, London, March 1987. The present paper was given there in an early version under the title "This Paradoxical Universe", and subsequently, in a different form, under the title "Probability and Determinism" at the May 1987 Vico Equense-Naples meeting of the Académie Internationale de Philosophie des Sciences. It was presented in yet fuller form under the title "Quantum as Foundation of Physics" at the Symposium on Basic Concepts in Quantum and Stochastic

Transport, IBM Thomas J. Watson Research Center, Yorktown Heights, New York, June 1987. The paper was further revised in content and title in September 1987 for publication in the IBM Journal of Research and Development; therefore it is Copyright 1988 by International Business Machines Corporation, reprinted with permission from the January 1988 issue of the IBM Journal of Research and Development. A single reference is listed here as point of access to some of the literature: J. A Wheeler, "How Come the Quantum", New Techniques and Ideas in Quantum Measurement Theory, Greenberger, D.M. (1987) Ed., Ann. New York Acad. Sci. 480: 304-316.

Valerio Tonini

A BRIEF NOTE ON THE RELATIONSHIP BETWEEN PROBABILITY, SELECTIVE STRATEGIES AND POSSIBLE MODELS.

Whenever a discussion arises about the concept of probability, the fact emerges that the verification of a possible event excludes everything that would have been possible if that event had not occurred.

The eternal debate concerning causality, probability and indeterminism would therefore be endless if a rigorous systematization of the information screening underlying scientific models had not clarified the selective strategies by which formally compact and consistent models are constructed; models which, even if they do not account for the innumerable variations of natural events, nevertheless permit an adequate representation of them within determinate limits of error and approximation.

From the historical point of view, the day of reckoning came when Louis de Broglie proposed a "causale et non linéaire" representation, by means of "la théorie de la double solution de la mécanique ondulatoire" so as to provide "une image du corpuscule où celui-ci apparaît comme le centre d'un phénomène ondulatoire étendu auquel il est intimement incorporé". But in a subsequent measuring operation, "il faudra construire un nouveau train d'onde de dimensions beaucoup plus réduites que celles du premier dans son état final. cette nouvelle forme de l'onde sera naturellement le point de départ d'une évolution des probabilités."

De Broglie, like Einstein, was dissatisfied with a conception of quantum mechanics which exacerbated the controversy Kontinuität versus Diskontinuät without solving it. It was just this question of the wave of probability proposed by de Broglie which induced me to write a note, which appeared in the March-April, 1948, number of "Scientia", entitled *Déterminisme*

E. Agazzi (ed.), Probability in the Sciences, 131–134.

et Indéterminisme, which focused on the crucial point in a dispute which had often reached a dramatic pitch in Einstein, Bohr and Born. It is unnecessary for me to recall either the crisis of Hilbert's systematics, provoked by Gödel's theorem, or Hermann Weyl's *Philosophy of Mathematics and Natural Science*, published in 1927, in order to arrive at what Husserl defined as the crisis of European science; a crisis which created paradoxes that always originated in a certain persistent dissension between the empirical mentality and logical formalism, between the reality of operational praxis and nomology. Such mutual incomprehension was fundamentally due to the fact that the sense of one irrefutable reality had been lost: that of the irreversibility of the real time in which any action occurs in a determinate structure.

The development of information theory in to a genuine theory of scientific knowledge has brought to light the fact that what is physically observable is the result of a complex informative operation within the field of a research project which continually reacts, historically and dialectically, with that aspect of reality which is being investigated. Hence the complexity challenge: it is now universally recognized that no single explicative system is sufficient to represent the structure of reality in all the innumerable varieties of its processes. Possible models of reality are the fruit of lengthy information strategies, necessarily selective, which can be simulated and represented logically and algorithmically, according tho theoretical paradigms that are formally compact, but different, even antithetical. Today these paradigms can be divided into four fundamental and contrastive categories: deterministic, probabilistic, indeterministic and cybernetic. No model constructed on the basis of one or another of these categories may be said to provide a descriptive and explicative procedure that is complete and exhaustive in itself; it can provide a more or less adequate interpretation and explanation of reality which is valid only in so far as it is the result of a well-controlled selective strategy. Today there are more sophisticated models concerned with neg-enthropic, metastable (Prigogine), catastrophic (Thom) and autopoeitic (Maturana, Varela) processes; but I maintain that they are fundamentally developments of one or the other of the four basic categories; consisting, in other words, in a complex re-combination of their reciprocal relationships. The important thing is to understand the limits of the explicative procedure within which one intends to work.

In physics, probability enters the play of the possible approximations and the so-called hidden variables; but in biology the problem has become more complicated since it was discovered that the structure of DNA constitutes the chemical support for genetic information, which guarantees the faithful reproduction of the species, with the possibility of combinations and changes which are the source of evolutionary processes. This discovery has given rise to a new, more wide-ranging conception of the biophysical and biopsychic world, because scientists have found that they are dealing, not so much with a complex combination of quantifiable facts, as with a system of processes, with innumerable "evolutionary fusions" that have ended by producing, among the numberless events the Universe, this act of human thought: an act which takes on itself the task of understanding the world, posing questions such as the following: is it possible to reconstruct and to prove, by means of astrophysical observation and microphysical calculations, a history of the Universe, starting with a supernatural explosion, and postulating in some way, at least statistically, the birth of organic forms?

Since the moment that this discovery was made, the theory of probability has undergone a complete change, because philosophical dualism of the Cartesian type, that is to say the idea of the heterogeneousness of the res cogitans with respect to the res extensa, seems to be less and less capable of explaining the complex and difficult relations between scientific thought and reality. And just as the contrast between the logical purity of "matematiche perfezioni" and the "imperfezioni della materia" (Galileo), was solved by means of the principle of conjugation in alternative models, so the need has now arisen for epistemologists to pay particular attention to what the biochemical, embriological and especially the neurological sciences are bringing out into relief with regard to the complex structuring process of the different factors that make up the human mind; since it is from the vital natural relationship which links every living thing to the ecosystem to which it belongs that the human mind draws the information that is the basis of all rational knowledge in all its various aspects: physiological, psychological, anthropological, cultural and social.

Once this high level has been reached in recognizing every aspect of the formation and development of the human mind, information screening is the tool of cybernetics, which is the science of the organization of neg-enthropic processes. Probability depends on the relationship between object structure and subject structure; and on whether these structures are endowed with memory. This excludes neither the unforeseen nor the improbable, nor the aleatory interaction of the so-called hidden variables. There are, therefore, different levels of information and different possibilities for cybernetic procedures. Science is the knowledge of real actions observed through a complex structure of information and checking. Hence the reappearance of the complexity challenge in information theory, that is, in the description of the way in which successive levels of knowledge are attained. At this point, a hermeneutic panorama unfolds: the panorama of all that cannot be formulated simply by means of one-to-one signs and logical systems, but by symbols having different interpretative values. The first science to be viewed in this new perspective should be psychology, now free both from its morbid, Freudian inheritance and from narrow-minded, materialist monism.

This brief note, summarizing a process that is far from linear, demonstrates how the concepts of probability and causality, like those of identity and otherness, together belong to that group of irrepressible antinomies which, as Kant pointed out, form an intrinsic part of human thought; the one cannot exist without the other.

PART 2

PROBABILITY, STATISTICS AND INFORMATION

Leslie Kish

CRITICAL REPLICATIONS FOR STATISTICAL DESIGN

While writing a new book [Wiley '87] I found several distinct statistical problems, each with a separate solution, but all in need of a higher principle of justification. I looked in vain into statistics for a higher principle or criterion which we need to serve as a common foundation for those similar solutions. Then I thought that "falsifiability" could be borrowed from the logic of scientific discovery, from the philosophy of science, if adopted in a suitably modified form to our needs in statistical design. However, discussions and correspondence with several philosophers of science and several statisticians suggested that I was wrong in trying to stretch falsifiability to cover our needs in statistics. I may also have been wrong in believing that falsifiability was both well known, well established, and well accepted as a logical principle of scientific discovery. Thus I am forced to propose a new name: *Critical Replication*. I can suggest several alternative names, some suggested by others, and ask for your preferences: sturdiness, robustness, resilience; sturdy conditioning, (re)conditioning, probing, replicability, generalizability, multiplicity.

Of course, more than names, a theoretical grounding for the new criterion or principle is needed. There let me merely point out several ways that such a principle of Critical Replication seems to differ from a strictly *logical* principle of falsifiability. (1) The principle must be frankly inductive or ampliative to be useful in statistical design. (2) Its statements must always be conditional, not categorical, in order to deal with the real worlds of empirical data. (3) It must be probabilistic, always based on uncertainties, never with the certainties of logical deduction. (4) It must be devoted chiefly to measuring errors and

E. Agazzi (ed.), Probability in the Sciences, 137–147.

limits, rather than searching for "exact" boundaries and classes. (But the "exact sciences" are concepts that statisticians never deal with - and scientists seldom in actual research.)

Statisticians, and scientists, for example, are not concerned with establishing the "existence" of gravitation, but to measure (with errors) the distribution of the strength of gravity on the earth's surface or of gravitation in the galaxy and universe. Also, we do not need to prove the existence of superconductivity, but to measure its properties as they are being newly revealed in 1987 and then to explain it.

Replication is a central aspect of all applications of statistics. It is at the core of all measurements of errors and of all aspects of statistical design. From that basis of replication it is a reasonable step to the principle of Critical Replication. I am unable to develop here the reliable and useful principles that are needed for Critical Replication, though a modest, brief attempt is made later. Rather permit me to share with you the nine examples I used for "Falsifiability in Statistical Design" in Section 7.6 of my new book on "Statistical Design in Research." Those nine examples do not exhaust the need for a higher principle for basic problems in statistical design.

Statistical design is a vital aspect of statistical research, yet design has been rather neglected in statistical theory, which tends to concentrate instead on the mathematical aspects of statistical analysis. My book aims to remedy in part this imbalance. Statistics and statistical design cannot avoid the basic philosophical problem of empirical science: to make inferences to large populations and infinite universes, to make broad and lasting generalizations from samples of data that are limited in scope and time; and which are also subject to random errors.

Since Hume in 1748, many philosophers, scientists and statisticians have written on this central problem of the philosophy of science, but the clearest and best known seemed to be Popper's view of "falsifiability."

"Now in my view there is no such thing as induction. Thus inference to theories, from singular statements which are 'verified by experience' (whatever that may mean), is logically inadmissible...Theories are, therefore, never empirically verifiable. But I shall certainly admit a system as empirical or scientific only if it is capable of being tested by experience. These considerations suggest that not the verifiability but the falsifiability of a system is to be taken as a criterion of demarcation. In other words: I shall not require of a scientific system that it shall be capable of being singled out, once and for all, in a positive sense; but I shall require that its logical form shall be such that it can be singled out, by means of empirical tests, in a negative sense: it must be possible for an empirical scientific system to be refuted by experience...My proposal is based upon an asymmetry between verifiability and falsifiability; an asymmetry which results from the logical form of universal statements. For these are never derivable from singular statements, but can be contradicted by singular statements." [Popper 1968, 6.1]. [See also Salmon, 1976]

"Popper's seminal achievement has been to offer an acceptable solution to the problem of induction. In doing this he has rejected the whole orthodox view of scientific method outlined so far in this chapter and replaced it with another. Popper's solution begins by pointing to a logical asymmetry between verification and falsification. To express it in terms of the logic of statements: although no number of observation statements reporting observations of white swans allows us logically to derive the universal statement 'All swans are white,' one single observation statement, reporting one single observation of a black swan, allows us logically to derive the statement 'Not all swans are white.' In this important logical sense empirical generalizations, though not verifiable, are falsifiable. This means than scientific laws are testable in spite of being unprovable: they can be tested by systematic attempts to refute them." [Magee 1973, pp. 222-23]

For statistical design of research our tasks are much tougher than separating white and black swans. Our "swans" come (metaphorically) in all shades of grey from white to black and often observed through a haze. "Statistics and statisticians deal with the effects of chance events on empirical data. Because chance, randomness, and error constitute the very core of statistics, we statisticians must include chance effects in our patterns, plans, designs, and inferences" [Kish 1982]. For example, statistics must distinguish the higher risks of death from lung cancer, heart attacks, etc., for smokers, although nonsmokers are also subject to those risks and many smokers die from other causes. Furthermore, many statistical problems of research are more complex than the effects of cigarette smoking.

Many scientists and statisticians, wrestling with problems of induction, have made statements resembling parts of Popper's falsification views. R. A. Fisher's espousal of multifactor design is an excellent example, and an independent contemporary of Popper's:

"We have seen that the factorial arrangement possesses two advantages over experiments involving only single factors: (i) Greater efficiency in that these factors are evaluated with the same precision by means of only a quarter of the number of observations that would otherwise be necessary; and (ii) Greater comprehensiveness in that, in addition to the 4 effects of single factors, their 11 possible interactions are evaluated. There is a third advantage which while less obvious than the former two, has an important bearing upon the utility of the experimental results in their practical application. This is that any conclusion, such as that it is advantageous to increase the quantity of a given ingredient, has a wider inductive basis when inferred from an experiment in which the quantities of other ingredients have been varied, than it would have from any amount of experimentation, in which these had been kept strictly constant. The exact standardization of experimental conditions, which is often thoughtlessly advocated as a panacea, always carries with it the real disadvantage that a highly standardized experiment supplies direct information only in respect to the narrow range of conditions achieved by standardization. Standardization, therefore, weakens rather than strengthens our ground for inferring a like result, when, as is invariably the case in practice, these conditions are somewhat varied." [Fisher 1935, p. 99]

But nowhere could I find examples of the practical problems of statistical designs described below. For these problems, the principles of falsifiability or of critical replication present a unified view that we need. These examples from the last Section 7.6 refer to earlier sections of my book, where these problems of design are discussed in detail. These earlier sections are noted in it after each example.

A. *Internal replication* (3.1B). If treatments and observations are introduced into several sites (communities) that are similar (adjacent) in resources, culture, organization, and administration, little additional *sei corroboration* is gained from the extra effort expanded on the replications. But if the replications are spread so as to increase (maximize) dissimilarities over the range of variables, over diverse conditions, over the entire signified populations of inference, then greater corroboration can be obtained from successful exposures to the severe tests of falsification, or critical replication. It is true that contradictory results from the several sites must lead to doubts and to the need for further research.

B. *Curves of response over time* (3.6). Many experiments should provide repeated observations for testing the effects of treatments over time, when effects may well vary over a long period, but more than mere short range effects are sought. Consider, for example, the very different effects on Romania's population growth, which fell drastically after the 1966 law abolishing easy abortion, but then rose again gradually as people rediscovered illegal abortions (Fig 3.6). Critical replications at intervals may guard against misleading optimism about responses to new treatments.

C. *Control strategies* (4.1). In observational studies, the difference $(y - y)$ between treatments A and B represents a theory about the source of that difference. But that theory must compete with others represented by uncontrolled disturbing variables, which may be confounded with the experimental treatments. Thus introducing statistical controls either in selection or in analysis (in order to test the treatment difference for survival against all those controls) represents tests of falsifiability, and surviving the most severe of the feasible tests leads to increasing corroboration for the proposed theory.

D. *Multifactor experimental designs.* R.A. Fisher's argument above for multifactor designs has basically much in common with those for multiple controls above. Yet it is even broader in that multifactor designs may begin with several factors and with interactions between them, in mutual competition for corroboration. The basic view of severe and broad tests against falsification is similar. Note that the bold view of this scientist-statistician predated those of the philosopher, but the two converge to the same basic point. Note that Fisher was also fighting against current views of induction, as Popper was. Fisher's ideas were further evolved in experimental design, especially in "response surface" analysis.

Remark. The views I am about to introduce pertain as much to examples A,B,C,D above as to E,F,G,H below. I fear to introduce this bold modification or relation of "falsifiability", which I have not found explicitly in the philosophical presentations. However, in daily, ordinary, "normal science" and in statistical design for research, we work less often with the "falsifiability" of bold theories than with "critical estimates" (to coin a term) for important variables. For example, we estimate *how much* cigarette smoking increases death rates from lung cancer and from heart failures, by what factor and what percentage. Or we estimate the dramatic convergence toward "zero population growth" of modern industrialized nations (both Western and Eastern Europe, North America, Australia, Japan, etc.) in our generation. This concentration on estimation of important parameters is common not only in social research, but also in statistics for the biological, chemical and physical sciences.

In statistics there is close correspondence between tests of significance and probability (confidence) intervals for estimates. Similarly, for statistical design a strong relation also exists between testing for "falsifiability" and "critical estimation" of important (crucial?) parameters. Many of the daily problems of statistical designs can be better understood in terms of estimation. Yet these can also be enriched with the falsifiability view. The problems of representation (1.1-1.4) enter here also, especially if we consider estimation not only for the entire population, but also for critical domains within it (2.3). Now we return to examples with this two-fold view in mind.

E. *Experiments, surveys, and controlled observations* (1.3). This is a rich area for critical replication, because usually the special strength of each method -randomized treatments, representation, and realism- cannot be combined into the same research design. For example, a randomized experiment of lifetime smoking/nonsmoking habits cannot be built into a randomized selection of the U.S. population. But the effects of cigarette smoking have been shown with all three methods, including animal experiments, human surveys, and controlled groups [Cornfield 1959]. Strong corroboration results from falsifying different competing causal theories with each method, whereas the smoking-cancer link withstands the varied tests of critical replication with all three methods. The Salk polio vaccine trial incorporated (uniquely) the features of all three methods in a single large study [Meier 1972]. Critical replication with different methods has other aspects as well. For example, changes in fertility behavior have been studied in the same population with both anthropological and with sample survey methods.

F. *Stratification.* This method of control is widely used in survey sampling to ensure good representation within the sample of identifiable subpopulations. Reducing variances appears as the chief justification in sampling theory, and representing domains for analysis (2.3) should also be considered. Controlling for potentially disturbing variables, although this is seldom made explicit, looms large especially for large, multipurpose samples and also when one does not know which of many potential stratifying variables may be the most important. Such situations lead to using several, even many, stratifiers and to "multiple stratification." [Kish and Anderson 1978; Kish 1965, 12.8]. This resembles the use of controls in C above, because without stratification, disturbing variables could have greater effects on the survey results. In experimental designs "blocking" serves similar purposes.

G. *Randomization of treatments* (1.1-1.4,3.2). This serves, when feasible, as the most powerful method for eliminating disturbing causal factors, which could compete with experimental treatments as predictors. Randomization provides severe tests against misleading results. After thorough investigations of social and medical innovations, Mosteller [1977] wrote with sarcasm "...that the less well controlled the study, the more enthusiasm the investigators have for the operation... nothing improves the performance of an innovation more than the lack of controls."

H. *Representation over the population* (1.1-1.4,3.1). Comparison of two methods, representative probability sampling versus internal replication (A), also illustrates the differences between tests of falsification and critical estimation. When the estimates are similar over all replications and over the entire population, both designs perform well; and internal replication may be more feasible, less costly, and more efficient. But when faced with different estimates, representative samples facilitate estimates for many kinds of domains - including those not covered by internal replications (2.3,3.1). They also yield global averages (over all domains) which are often needed, but that a few internal replications cannot yield.

I. *Publication of contradictory results.* "On rare occasions a journal can publish two research papers back-to-back, each appearing quite sound in itself, that comes to conclusions that are incomparable in whole or in part. Such a conjunction can put critical issues involving research methods and interpretations in unusually sharp focus. I recall that the *Journal* published such papers in 1978 and 1982, each with a helpful editorial. In this issue we have another such pair -and both appear to me methodologically sound. Each cites prior studies that support its conclusions and others that do not; neither is alone". [Bailar 1985]. I greet the wisdom and courage of this Journal for printing both articles, but some believe it is embarrassing for science and confusing for the public to reveal the contradiction. Should the journal wait for further evidence? But checks have already found both reports to be methodologically sound; and each is already both supported and contradicted by other studies.

Would it be better to have the two results appear in two different journals? In two different countries, (like the USA and Cuba)? On two different continents; say one in Japan, or Indonesia, or Egypt? That would allow for speculations about sources of human diversity. Such differences often allow conflicting results in human and social research to coexist without clear contradiction. I believe that such contradictions are productive results of critical replications that lead to stronger theories -to be tested anew.

Replications, errors, and chance are concepts linked intimately to statistics and to each other. When dealing with empirical data from *any* field, the separation of chance effects from causal relations becomes the function of statistics. [Kish 1982]. Here we also have common grounds between statistics and philosophy for all the empirical sciences. Statistical measurements of errors depend on replications, but encounter philosophical obstacles, as a philosophical statistician wrote me in a friendly letter:

"Of course I agree that replication is centrally important...Nonetheless, I think there are major conceptual difficulties with the replication concept...The main point is that you can never step into the same river twice, and Lucretius might have gone a bit further and said that you can never exactly define the river at a specific time. In short, replications under the same circumstances are impossible...Since true replication is impossible, how can the concept be so central?"

We discuss this question in its statistical aspects, because all statistics is based on replications. But the philosophical problem of classification has other scientific aspects also, because empirical science cannot be based on unreplicable, unique (historical) phenomena. Going beyond the river of Lucretius, think of making replicate measurements on any of the following magnitude:

a) The area of the ozone hole over the South Pole.
b) The number of red blood cells in all your blood or in 1 cc of it.
c) The population of Chicago.
d) The number of unemployed in the USA.

While you read that paragraph all of those magnitudes have changed. Therefore replicate measurements taken at the two ends of that brief period would contain both strictly "measurement" (instrumental) errors" (including the effects of the measurements themselves) and changes in the subjects (populations). Thus these "chance (or random) replications" contain both strictly instrumental errors and also basic chance variations, the chance effects of the objective world beyond the observer's control.

These objective, chance effects become greater as the time between the replicates is increased - to an hour, day, week, to a year. The spread between the measurements may take many forms beyond the passage of time; for example, consider the average of red cells in US persons of your age and sex; then of all ages and sexes; then of people from other countries also. As we broaden the populations of inference we move from simple chance replications to critical replications. There is no clear, unique boundary between these two kinds of replications and the demarcation depends on the inferential aims of the study.

Furthermore, the different types of events, of phenomena (of different years, different ages, sexes, different countries) may become the domains, objects of the study. Then we study them in causal relations rather than as critical replications to be included within a single population of the research. Again the boundaries between the two types are not unique, but must be established by the framework of the research.

Thus Critical Replications must be distinguished from simple chance (random) replications on one side; and from domains of causal relations on the other. In both senses they become important aspects of statistical design. Experimental designs distinguish "random" from "fixed" effects (models) and this dichotomy resembles somewhat the distinction of chance replications from causal domains above. However, I know of no discussion of Critical Replications in our sense, except for "stratification" and "blocking", as in F above. There was "falsifiability" in philosophy with the disadvantages noted above.

The University of Michigan (U.S.A.)

BIBLIOGRAPHY

Bailar, J. C. (1985), When research results conflict, New England Journal of Medicine 313: 1808-81.

Cornfield, J., Haenszel, W., Hammond, E. C., Lilienfeld, A. M., Shimkin, M. B., Wynder, E. L., et al. (1959), Smoking and lung cancer, Journal of the National Cancer Institute 22: 173-203.

Fisher, R. A. (1935), The Design of Experiments. 1st ed. London: Oliver and Boyd.

Kish, L.:

(1965), Survey Sampling. New York: John Wiley and Sons.

(1982), Chance, Statistics, Sampling. Statistica 42: 3-20. Reprinted in Journal of Official Statistics. (1985), vol 1: 35-47.

(1987), Statistical Design for Research. New York: John Wiley and Sons.

Magee, B. (1973), Popper. London: Fontana.

Meier, P. (1972), "The biggest public health experiment ever: the 1954 field trial of the Salk polio vaccine". In Statistics: Guide to the Unknown, edited by Tanur, J. M. et al.: San Francisco: Holden-Day.

Mosteller, F. (1977), Experimentation and innovations. Bulletin of the International Statistical Institute 41st Session.

Popper, K. R. (1959), The Logic of Scientific Discovery. London: Hutchinson (esp. Chs. 1, 10).

Salmon, W. C. (1967), The Foundation of Scientific Inference. Pittsburgh: University of Pittsburgh Press (esp. Secs. 1, 7).

Mario Casartelli

THE CONTRIBUTION OF A.N. KOLMOGOROV TO THE NOTION OF ENTROPY

The immediate source of the concept of *entropy of a dynamical system*, introduced by A. N. Kolmogorov in 1958, was a definite and typical problem in the context of abstract dynamical systems, the problem of *isomorphism*. In spite of being so natural, it is far from being immediately evident or soluble. Consider, for instance, a classical case of probabilistic games (mathematically: Bernoulli Shifts): it is quite obvious that, in a repeated experiment, tossing a coin is equivalent to tossing a die and considering the outcomes only in terms of whether they are odd or even; should it also be equivalent to tossing a die and conceiving of the outcomes as ranging from 1 to 6? It was by using his new 'metric invariant' - the dynamical or K-entropy - that Kolmogorov succeeded in giving a negative answer to this old and only apparently simple problem.

Nowadays the notion of Dynamical Entropy is situated at the crossroads of many lines of research in mathematics and physics, and affords them a unifying point. To appreciate its conceptual fertility, some formal definitions are needed. This involves what is essentially a three-step process, defining, in order, I: $H(\alpha)$, the entropy of a partition α, II: $H(T, \alpha)$, the entropy of a transformation T with respect to a partition α, and finally, III: $h(T)$, the dynamical entropy of the transformation T.

I

To define $H(\alpha)$, the entropy of a partition α, we shall first consider Shannon's entropy (1948), namely the notion of entropy

E. Agazzi (ed.), Probability in the Sciences, 149–159.

of a finite experiment. According to Shannon's conception, in a space $\mathbf{M}$ with a probability measure μ defined on a σ-algebra $\mathcal{M}$ of subsets, a finite partition α is a collection of pairwise disjoint measurable sets $A_1, A_2, ..., A_N$ (also called *cells* or *atoms* of the partition), whose union is the whole of $\mathbf{M}$. Its entropy $H(\alpha)$ is then

$$H(\alpha) = -\sum_{k=1}^{N} \mu(A_k) \log \mu(A_k) \tag{1}$$

This formula shows that $H(\alpha)$ coincides with the Shannon entropy of an experiment whose single events, with their probabilities, are represented by atoms A_k and their measures. Henceforth we shall speak indifferently of the *entropy of an experiment* and the *entropy of a partition*. The letters of the alphabet as they appear in a text, seen as a sequence of outcomes from a probabilistic source, constitute a good and typical example of repeated experiments.

Connections with Information Theory would constitute, in principle, a highly problematic field; Kolmogorov has in his later works (1965, 1969) turned to the basic definitions of information, where he has made remarkable contributions. His celebrated notion of *algorithmic complexity*, in particular, took its beginning directly from a discussion on various possible approaches to a definition of information (combinatorial, probabilistic etc.) But with regard to our present needs, definition (1) may be more simply justified by intuitive arguments: the quantity of information a person has when he knows a single outcome of an experiment is assumed to be equal to the minimum length in symbols of a complete description of this outcome (complete, of course, in the sense of sufficing to distinguish it from other possible outcomes). This length is proportional to the logarithm of the inverse probability of the event. The basis of the logarithm fixes the unit (e.g. basis 2 measures information in *bits*). The plausibility of such an assumption clearly appears from the elementary case of identifying a string of N binary symbols: there are 2^N possible strings, each of them with a probability $1/2^N$. The relation $N = -\log_2(1/2^N)$

indeed gives the length of the description in bits. The expectation of the outcome of the experiment as a whole, i.e. the weighted average of the informations had in knowing single outcomes (also a measure of the ignorance or uncertainty removed) coincides with the entropy (1). A strictly connected and important concept is the *conditional* entropy of α with respect to β, defined as:

$$H(\alpha|\beta) = -\sum_{i,j} \mu(A_i \cap B_j) \log \mu(A_i|B_j) \tag{2}$$

The above means: the residual uncertainty about the outcome of experiment α when the result of β is known. This may be read in the fundamental relation:

$$H(\alpha|\beta) = H(\alpha \vee \beta) - H(\beta) \tag{3}$$

The symbol "$\vee$" denotes a sort of product between partitions: $\gamma = \alpha \vee \beta$ is the minimal partition containing both α and β.

Khinchin proved the following uniqueness theorem: If a functional $f(\alpha) = f(\mu(A_1), ..., \mu(A_N))$ enjoys the following properties:
i) f is continuous for any N with respect to all its arguments;
ii) for fixed α, f takes its maximum value when $\mu(A_k) = 1/N$;
iii) $f(\alpha|\beta) = f(\alpha \vee \beta) - f(\beta)$; and
iv) $\alpha \equiv \alpha'$ up to cells of measure 0 $\Rightarrow f(\alpha) = f(\alpha')$,
then

$$f(\mu(A_1), ..., \mu(A_N)) = -\lambda \sum_{k=1}^{N} \mu(A_k) \log \mu(A_k)$$

i.e. f coincides with the entropy (1) up to a multiplicative constant.

But why "entropy"? There are both historical and formal reasons for this. We note that Kolmogorov, assuming the Shannon entropy (1) as the point of departure for his new notion, implicitly inherited the whole set of technical and philosophical arguments regarding the relations between entropies from various fields (thermodynamics, statistics, information theory). At this level, his most original contributions emerged later, as mentioned above. Khinchin's theorem shows that $H(\alpha)$ satisfies certain general qualitative requirements of an index of "disorder".

II

$H(\alpha)$ is only the first step. K-entropy differs from Shannon's entropy in that it does not deal with information from experiments (or partitions) but with information from *dynamics*. One could observe that in speaking about the entropy of a source of messages - his reference model - Shannon also introduces a dynamics. But the time evolution of the source may be reduced in such a case to a single type of dynamics, the so called "shift", i.e. a translation in a space of sequences. The Kolmogorov notion of entropy applies to *every* measure preserving transformation T, defining in this way a functional which is meaningful in the general context of dynamical systems, with implications for instance in ergodic theory, dimension theory and the study of the onset of stochasticity.

For present purposes we may omit a discussion of problems arising from the possible noninvertibility of T; moreover we shall limit ourselves to considering a discountinuous time (i.e. an orbit is given by $x, Tx, T^2x, T^3x, \ldots$). The evolution of a partition is also introduced quite naturally.

The next step consists therefore in defining $H(T, \alpha)$, the entropy of T with respect to α. There are two ways of doing this, which, even if mathematically equivalent, stress the meaning of the new quantity differently. The first approach gives:

$$H(T, \alpha) = \lim_{N \to \infty} \frac{1}{(N+1)} H(\alpha \vee T\alpha \vee \ldots \vee T^N \alpha) \qquad (4)$$

and the second

$$H(T, \alpha) = \lim_{N \to \infty} H(\alpha | T^{-1}\alpha \vee \ldots \vee T^{-N}\alpha) \qquad (5)$$

Consider first definition (4): the argument of H is a growing sequence of partitions, a 'composed experiment' obtained through the product of the evolving $T^k\alpha$. It is almost obvious that entropy will increase with increasing partitions. One may expect that a stabilized partition would arise asymptotically, and therefore that its entropy, once divided by $(N+1)$, necessarily goes to 0 in the limit. This happens indeed in several cases, for instance, when α is periodic under T (i.e. when $T^M\alpha = \alpha$ for some M): the maximal argument of H in the r.h.s. of (4), in such a case, is $\bigvee_{k=0}^{M} T^k\alpha$, and $H(T,\alpha)$ is trivially 0. Actually, for certain T there exist partitions in which the limit is different from 0, and this fact entails a remarkable property: the evolution induced by T is so complicated that the entropy in the r.h.s. grows with N, i.e. is proportional to the time.

The same mathematical content appears in a different form in (5), where $H(T,\alpha)$ is given as the residual uncertainty about α when the past history of its evolution is known. It characterizes T with regard to its ability to produce in the observer the necessity of acquiring new information, independently of the past. Now the possibility of positive entropy clearly appears (at least qualitatively) when we consider for instance that the past history of an 'honest' roulette wheel does not give any information about the next outcome. In this sense, formula (5) is more explicit than (4), and may also be read in the following way: positiveness of entropy is equivalent to finiteness of memory.

III

Finally, we define:

$$h(T) = \sup_{\alpha} H(T,\alpha) \tag{6}$$

The sup operation eliminates the dependence on partitions, defining entropy through the extremal possibility of T in demanding new information.

Some facts should be noted:

1. The above definition (essentially due to Sinai, 1959) was originally given by Kolmogorov in a slightly different form.
2. If there are several measures $\mu, \nu, \ldots$ on the same algebra $\mathcal{M}$, to each measure will correspond a distinct entropy $h_\mu(T), h_\nu(T), \ldots$
3. The triple $(\mathbf{M}, \mathcal{M}, \mu)$ properly speaking is a measured space where μ has to be intended as 'probability' only in an axiomatic sense, without any reference to an intrinsic *probabilistic* as opposed to *deterministic* feature of the system. For example, the phase space of a classical system with the Liouville measure fulfills this requirement.
4. The role of μ is redundant in another sense: the construction of Kolmogorov and Sinai introduces a functional of T whose qualitative meaning (entropy as an index of 'disorder') does not strictly depend on a probabilistic framework. In successive developments of the theory (1965) Adler, Konheim and McAndrew succeeded indeed in extending the concept of entropy in topological spaces, where a probability measure is not necessarily defined. The only difference with respect to the just defined 'metric' quantity appears in the point of departure (1), where, instead of partitions, one speaks of *coverings*: if $\mathbf{M}$ is a compact topological space, for every covering α there exists a finite number $n(\alpha)$, the cardinality of the smallest finite subcovering: the topological entropy of α, $H^{top}(\alpha)$, is $\log n(\alpha)$. Then, formulas (2) to (6) may be used to define the topological entropy $h^{top}(T)$. An important result is that, for topological spaces endowed with a probability structure, $h^{top}(T) = \sup h_\mu(T)$, the sup being taken over all T-invariant measures.
5. Entropy $h(T)$ is invariant under isomorphism, i.e. if there exists a one-to-one correspondence $\Phi : (\mathbf{M}, \mathcal{M}, \mu, T) \to (\mathbf{M}', \mathcal{M}', \mu', T')$, then $h(T) = h(T')$. To prove that two systems are not isomorphic, it is therefore sufficient to prove that they have different entropy. It was on the basis of such considerations that Kolmogorov succeeded in solving the old problem of classifying automorphisms with identical spectral type into distinct classes of isomorphism: tossing a coin or a die are *not* isomorphic processes, since respectively have

$h(T) = \log(2)$ and $\log(6)$. Nevertheless, $h(T)$ fails in general to be a *complete* invariant, i.e. a quantity such that, when equal for two systems, ensures that they are isomorphic. (D. Ornstein has proved that K-entropy is a complete invariant for Bernoulli Shifts.)
6. An operative difficulty would arise if one had to compute $h(T)$ directly from definition (6): the sup over all measurable partitions is indeed, in principle, a difficult operation to perform. Important developments of the theory are therefore related to the possibility of implementing (6) with an effective computational procedure. In this context two results seem to be particularly important: the first, due to Kolmogorov and Sinai, singles out a class of partitions ('generators') maximizing $H(T, \alpha)$ (i.e. if γ is a generator, $h(T) = H(T, \gamma)$; this concept was already present in the early work of Kolmogorov); the second is a set of contributions by various authors which, on the basis of results by Pesin and Oseledec, relates the entropy of dynamical systems to such computable observables as Lyapunov exponents.

This last point should be seen in connection with the observation in remark 3. Can a 'deterministic' system have positive entropy? The question could appear to be a simple matter of definition, since there are authors (for instance W. Parry or D. Ornstein) who propose defining determinism for dynamical systems precisely as the property of having zero entropy. But the notion of determinism is prior to and independent of that of entropy: for instance, we normally call those systems where the evolution of dynamical observables is ruled by differential equations deterministic. Hamiltonian systems constitute paradigmatic examples. Then, for definiteness, one could ask: can a Hamiltonian system have positive entropy? The answer 'yes' would imply this non-trivial consequence: notwithstanding the fact that for every Hamiltonian system and for every set of initial conditions a complete knowledge of the present virtually contains the whole of the future (and this definitely distinguishes them from intrinsically probabilistic systems, such as roulette wheels) there exists the possibility that a Hamiltonian system continuously reproduces, in its evolution, a new situation of ignorance in the observer, exactly as a probabilistic system does.

To convince oneself of this possibility, consider an ε-partition of the accessible region of the phase space of a Hamiltonian system (the energy surface Σ_E), i.e. a partition into cells whose width is at most ε. An orbit is then characterized up to an ε precision-level by successive occupations of cells on Σ_E. As $\varepsilon \to 0$, this description goes to the continuum description of the ideal orbit. But, before going to the limit, consider two orbits starting from the same cell. There are essentially two possibilities: 1) the fact that orbits have started very close to one another is sufficient to guarantee that they remain close at any time; 2) after a certain time, the initial closeness becomes irrelevant, and the two orbits evolve independently. Moreover, suppose that this happens independently of ε, or, in other words, that inasmuch as two orbits are *distinct* they are also *independent*. Analysis shows that the answer to the question 'in which cell lies the orbit at time t?' requires, in case 2, a continuous supply of information, or equivalently that entropy be positive (while it be 0 in the former case). Yet determinism is not under discussion. Of course, similar considerations apply to many deterministic models other than the Hamiltonian systems.

It is an interesting point that the possibility outlined above is not merely theoretical, but may be illustrated by rigorous proofs (in a few cases) and by numerical experiments (in many cases). As P. Grassberger has said "One of the most fascinating developments of the last two decades is the enormously rich structure which is observed in very simple dynamical systems" (he is here referring to logistic maps, Lorenz equations, Bernard cells, nonlinear driven oscillators, cellular automata, and so on). And, as D. Ornstein has observed, considering "geodesic flow, billiard or hard sphere gas we see that even though the laws governing the time evolution of these systems are completely deterministic, as soon as we make a measurement with only a finite number of possible outcomes (watch it on a TV screen for example) we get a process that is essentially indistinguishable from a process with finite memory." (Incidentally, we note here that Ornstein must use the expression *deterministic laws* to escape his internal convention according to which only the systems with zero entropy are deterministic). Lyapunov exponents may be recalled here: they indeed characterize the degree of independence of orbits, measuring their rate of local divergence.

These ideas, connecting independence of orbits, the having of positive entropy, sensitivity to initial conditions and so on, have to be correctly referred to a definite level of description. Randomness and stability, which are opposed at the same level, may coexist and support one another when referred to distinct levels. For a typical case of possible misunderstanding on this point, we recall that when the Ergodic Theory started with Boltzmann the idea was to legitimate the use of Gibbs statistical ensembles, thereby supporting, through a rigorous mathematical formulation, the independence of the macroscopic behaviour and features of physical systems from such microscopic details as "initial conditions" in the phase space. Now, as shown above, this claim may be considered satisfied when orbits *strongly depend* on initial conditions, or, in other words: the greater the dependence on the initial conditions of *orbits*, the less the dependence of the *system* as a whole. On the contrary, integrable and near-integrable systems (systems with 0 entropy, nonergodic systems, and so on), which are sensitive to initial conditions and do not allow the use of Gibbs ensembles, are those whose orbits are stable. Such an exchange of roles between randomness-sensitivity and stability is quite natural and familiar in other fields: it seems obvious, for instance, that the value of a statistical observable becomes more stable and reliable as the sample becomes larger. In in a sense, information about the ensemble is ensured by uncertainty about individuals.

Of course, these considerations open a sequence of related questions: for instance, what does "independence" or "randomness" really mean? Another example (taken from Grassberger, but going back in its essential lines to Kolmogorov himself) can clarify these points. Consider a sequence that both common sense and rigorous mathematics would call "random", the sequence of decimal digits in π. There are exact formulas giving π as a series, so that there are algorithms that in principle can produce as many digits as one wants. Then, what does it mean to say that the 'next' digit is random with respect to all the preceding digits? The fact is that if one considers not the first N digits, but a *string* of N digits of π in *any* position (requiring, in other words, a sort of translational invariance), then the shortest way to specify the string consists in listing it: and the length of the list grows with N, independently of the previous digits. Here the idea of *algorithmic complexity* (determining the randomness of a sequence on the basis of the length of the program producing it) rejoins the notion of entropy introduced above. In other words, irrational numbers or stochastic deterministic systems are *actually infinite* sources of information (we say "infinite sources of information" and not "sources of infinite information" because this latter definition should more properly be reserved for processes with infinite entropy, such as Brownian motion for instance).

We have seen that in the present context it is possible to translate questions concerning deterministic systems into questions related to the existence of the continuum, a fact which may have implications regarding the philosophical aspects of the foundations of analysis and dynamics which, as far as we know, have yet to be fully developed.

University of Parma (Italy)

BIBLIOGRAPHY

Adler, R.L., Konheim, A.G., and McAndrew, M.H. 'Topological Entropy', *Trans. Amer. Math. Soc.* **114** (1965), p.309.

Bai-lin, Hao, *Chaos*, World Scientific, 1984.

Billingsley, P. *Ergodic Theory and Information*, New York: John Wiley, 1965.

Grassberger, P. 'Self-Generated Complexity', *Statphys* **16**, H.E. Stanley (ed.), 1986.

Khinchin, A.J., *Mathematical Foundations of Information Theory*, New York: Dover, 1957.

Kolmogorov, A.N. 'A New Metric Invariant of Transient Dynamical Systems', *Dokl. Akad. Nauk.* **119** (1958), p. 861.

Kolmogorov, A.N., 'On Entropy per Unit Time as a Metric Invariant of Automorphism', *Dokl. Akad. Nauk.* **124** (1959), p. 754.

Kolmogorov, A.N., 'Three Approaches to the Quantitative Definition of Information', *Problemy Peredachi Informatsii* **1** (1965), no. 1, p. 3.

Kolmogorov, A.N., 'On the Logical Foundations of Information Theory and Probability Theory', *Problemy Peredachi Informatsii* **5** (1969), no. 3, p. 3.

Lichtenberg, A.J. and Liebermann, M.A., *Regular and Stochastic Motion*, Berlin, Heidelberg, New York: Springer, 1983.

Martin-Löf, P., 'The Definition of Random Sequences', *Information and Control* **9** (1966), p. 602.

Ornstein, D.S., 'Bernoulli Shifts with the Same Entropy are Isomorphic', *Adavances in Math.* **4** (1970), p. 337.

Ornstein, D.S., 'What Does it Mean for a Mechanical System to be Isomorphic to the Bernoulli Flow?' *Dynamical Systems, Theory and Applications*, J. Moser (ed.), Berlin, Heidelberg, New York: Springer, 1975.

Parry, W., *Entropy and Generators in Ergodic Theory*, New York: Benjamin, 1969.

Rokhlin, V.A., 'Lectures on the Entropy Theory of Measure Preserving Transformation', *Russian Math. Survey* **22** (1967).

Sinai,Ya., 'On the Concept of Entropy for Dynamical Systems', *Dokl. Akad. Nauk.* **124** (1959), p. 768.

Sinai, Ya., *Introduction to Ergodic Theory*, Princeton: Princeton University Press, 1976.

Hans Freudenthal

THE PROBABILITY OF SINGULAR EVENTS

Let me begin by directly asking the question: What is the probability that within the next year (starting April, 1987) a treaty on nuclear disarmament will be signed between the United States and the Soviet Union?

I am sure that your answer will be that the probability is close to one.

If, on the other hand, I ask you the probability of a treaty being signed during the same period which totally abolishes nuclear weapons, I expect your answer will be that it is close to zero.

And if I qualify my original question to varying degrees, the answers I receive will vary between zero and one. But rather than ask such questions, I shall pose a more fundamental one: What kind of probability is being referred to here? Is it the same kind as where one states the probability of a die's showing a six, or of a radioactive atom's disintegrating within the next second? Is it the same kind of probability as that of a haemophiliac person's having a grandson with the same defect, the same kind one attributes to weather forecasts, to the correctness of predictions of the outcomes of football games, of the ups and downs of the dollar exchange rate, of political events?

The line I have staked out with these examples shows a clear trend. It is a gradual trend which suggests that there is a frontier beyond which one should refrain from using the term "probability". But if this is so, where should it be drawn, and what would be the criteria for admitting one use and rejecting another?

This uncertainty has been the source of the distinction between the concepts of *objective* and *subjective* probability, as

E. Agazzi (ed.), Probability in the Sciences, 161–164.

well as of quarrels and controversies, which to my mind are pointless symptoms of dogmatism. I would prefer rather a pragmatic attitude. The habit of applying the terminology of chance and probability in quite disparate situations is a fact to be respected; none has authorised mathematicians or logicians to veto one use or the other. On the contrary I think a better policy is to be responsive, certainly if one's aims is to clarify the logical state of probability terminology.

One can *measure* lengths, weights, and time intervals; or, lacking appropriate standards and precise procedures, one can satisfy oneself with *estimating* them. But let us make it clear that these are only gradual differences. Even the most precise measurement is still an estimation as long as one is not absolutely sure of the last decimal. Once upon a time the radius of the universe was estimated, only to have this value substantially corrected later on. At any one instant one can avail oneself of all the available information, but the treasure of information, empirical and theoretical, grows continuously.

Does it make sense to speak of *subjective* lengths, weights and time measures? Certainly not. If understood in the same sense, *subjective* probabilities do not differ from *objective* ones in principle. I would rather avoid these misleading terms. The idea of probability is the same in the natural sciences, in the social sciences, in the activities of the researcher, the merchant, the lawyer, the politician and the man in the street. What is different, is the degree of accuracy in estimating and calculating probabilities.

Estimations and calculations need not be made explicit. Often people's behaviour in conditions of uncertainty betrays their ideas concerning in particular probabilities. The most explicit behaviour is betting. Laying p to q that something will happen means estimating the probabilities in the same ratio. It is a subjective estimation; however, it is one which can be stabilised at the market where bets are offered and accepted, so as to express intersubjectivity. Such estimations are based on personal experience and more or less complete information. Many people are habitual betters, and the sequence of their successes and failures may even lead to a frequentist interpretation of their future chances of winning or losing.

Implicit or even unconscious bets are even more frequent. The stock exchange quotations, and in particular the prices of options, reflect probabilities, estimated not solely with respect to particular stocks, but even more to such general economical and political developments as I have chosen to formulate in my first questions.

Nuclear disarmament would be a singular event. Viewed in a certain perspective, every event is singular. Births, deaths - these are singular events in the eyes of those most affected by them; but they are less singular when viewed by midwives, physicians, undertakers and registry officers, in whose hands they eventually lose all their singularity through becoming part of the population statistics.

Even every throw of a die, every disintegration of a radioactive atom is a singular phenomenon because of the individuality of the die or atom concerned and the specificity of the moment and place where these events happen, although this singularity is lost in the anonymity prevailing in the large families of similar events to which they belong.

Indeed, it is with respect to large families of similar events that probabilities are calculated with the greatest precision. The life expectancy of a man of forty in a given country is well defined, but as soon as supplementary information is provided concerning the man, this expectancy will change and its precision diminish. The family of events to which this man's life expectancy belongs will shrink as more information is added, until it consists of his life expectancy alone - which still will not tell us the date of his death. Nevertheless, insurance companies are prepared to underwrite policies on such singularities as the demise of a film star's beautiful legs, which means that they are able or believe themselves able to estimate the probabilities concerned.

Let us reconsider our example of a treaty on nuclear disarmament. Due to its many ramifications it can exercise more or less sweeping influences on the weapon industry, and indirectly on the economy as a whole. It is the concern of practical economists to anticipate future developments in order to propose specific measures and to develop strategies of action. Rest assured that at the various institutes of military and business research, computer models are being designed to simulate worlds where the state of nuclear defence differs from the present one in various ways, and the political and economic consequences of these scenario are calculated. No doubt these models involve parameters which are nothing other than probabilities of absolutely singular events, the values of which are estimated and revised on the strength of all the new information available. Rest also assured that already now many political and economic actions are being influenced by research of this sort, which in turn influences the estimation of future probabilities.

I have for a long time regretted that philosophers and methodologists of science pay too little attention to what actually happens in science and its applications to the real world. Often their theories reflect only a caricature of scientific reality. The quarrel over subjective and objective probabilities is a particularly striking example of this attitude.

Vakgroep OW & OC (The Netherlands)

Arturo Carsetti

PROBABILITY, RANDOMNESS AND INFORMATION

1. Let us represent the successive outcomes of an experiment by a sequence of independent and identically distributed random variables on some probability space. Let us assume that these random variables will be binary valued taking the values 0 and 1. As it is well known, according to R. von Mises' definition of randomness, the first condition for the existence of a random sequence (or Kollectiv) is that the limit of the relative frequencies of 0's and 1's exist; the second requirement is that the limit frequence should remain invariant under place selection made by any countable set of place selection functions.

As. J. Ville showed in 1939 these two requirements are not sufficient to obtain a correct definition of randomness. Actually it is possible to construct a Kollectiv according to von Mises' conditions which does not satisfy the law of the iterated logarithm. In order to identify a more adequate definition of randomness, capable of overcoming Ville's objection, two major strategies can be pursued: the algorithmic approach due to Kolmogorov and Chaitin and the measure-theoretic approach outlined by Martin Löf. These different paths of research lead to two distinct definitions of random sequence that, as we shall see, can be considered, within certain limits, as equivalents.

The fundamental idea of Martin Löf is that the probability laws which hold with probability one or "almost everywhere" such as the strong law of large numbers or the law of the iterated logarithm and which can be defined and tested by effective methods, are the same as those defined by the (total) recursive sequential tests. Actually, generally speaking, we can take an hypothesis and represent it by a probability measure on infinite sequences. The given sequence, if it is random, should satisfy theorems of probability based on that hypothesis. On each theorem we can base a test. Obviously we cannot decide in

E. Agazzi (ed.), Probability in the Sciences, 165–172.

a finite number of steps whether or not an infinite sequence x is random: the decision must be made on the basis of a finite segment only. Hence the necessity to make the tests sequential: once an initial segment of a sequence has been rejected as random, all extensions of it are also rejected as random at that level. Let X^* be the set of all finite sequences of 0's an 1's. Let X^∞ be the set of all infinite sequences of 0's and 1's. Let m denote the induced product measure on X^∞ and let $[A] = A\ X^\infty$, (for all $A \in X^*$).

Definition 1

A recursive sequential test is a recursively enumerable set $U \subset X^* \times N$ with:

$$.U_i = \{ x \mid (x,i) \in U \}.$$

$$m\,[U_i] \leq 2^{-i}$$

$Q_U \subset \bigcap_{i \in N} [U_i]$, is a recursive null set with respect to m.

The sequence $\{U_i\}_{i \in N}$ approximates constructively a set $Q_U \subset X^\infty$ of m-measure zero uniquely associated with a probability law which holds with probability one or almost everywhere.

Definition 2

A recursive sequential test U is a total recursive sequential test if $m\,[U_i]$ is computable for all $i \in N$.

On the basis of these definitions, we can say that a sequence $x \in X^\infty$ is random, according to Martin Löf's conception of randomness, if x does not belong to Q_U for all recursive sequential tests U.

2. The starting point of the algorithmic approach to the definition of randomness has an informational character. Classical information theory says that messages from an information source that is not completely random can be compressed. In this sense the definition of randomness is merely the converse of this fundamental theorem of information theory: if lack of ran-

domness in a message allows it to be coded into a shorter sequence, then the random messages must be those that cannot be coded into shorter messages.

According to this general idea we can distinguish the three concepts of algorithmic probability, algorithmic entropy and algorithmic complexity. Let us consider a Turing Machine Q with three tapes (a program tape, a work tape and an output tape) and with a finite number n of states. Q is defined by an $n \times 3$ table and equipped with an oracle (we omit, of course, all the details). For each TM Q of this type $P(s)$ is the probability that Q eventually halts with the string s written on its output tape if each square of the program tape is filled with a 0 or a 1 by a separate toss of an unbiased coin. $H(s) = -\log_2 P(s)$ will be the algorithmic entropy. Finally, the complexity $I(s)$ is defined to be the least n such that for some contents of its program tape, Q eventually halts with s written on the output tape after reading precisely n squares of the program tape. In other words I is the minimum number of bits required to specify an algorithm for Q to calculate s. From the concept of complexity we can derive, as usual, the concept of conditional complexity: $I(s/K_1...K_n)$.

Definition 3

A string s is random, according to the algorithmic approach, iff $H(s)$ is approximately equal to $|s| + H(|s|)$. An infinite string z is random iff there is a c such that $H(z_{\underline{n}}) > n - c$ for all n, where $z_{\underline{n}}$ denotes the first n bits of z.

Definition 4

The base-two representation of the probability Ω that the universal computer R halts is a random infinite string: $\Omega = \sum. 2^{-|p|}$.

In other words, we can say that the basic theorem of recursive function theory that the halting problem is unsolvable corresponds in algorithmic information theory to the theorem that the probability of halting is algorithmically random if the program p is chosen by coin flipping. We can also remark that the above definition of randomness is equivalent to saying that the most random strings of length n have $H(s/n)$ close to n, while the least random ones have $H(s/n)$ close to 0.

Finally we have to stress that, even though most strings are algorithmically random, an inherent limitation of formal axiomatic theories is that a lower bound n on the entropy of a specific string can be established only if n is less than the entropy of the axioms of the formal theory.

3. C.P. Schnorr (1974) and R.M. Solovay (1975) have shown that the complexity-based definition of a random infinite string and Martin Löf's statistical definition of the same concept are equivalent. Generalizing these results G. Chaitin (1987) has shown that the Martin Löf's definition of a random real number using constructive measure theory is equivalent to the complexity-based definition of an algorithmically infinite random string x or real x.

We remember that speaking geometrically a real r is Martin Löf random if it is never the case that it is contained in each set of an r.e. infinite sequence A_i of sets of intervals with the property that the measure of the ith set is always less than or equal to 2^{-i}:

$$m(A_i) \leq 2^{-i}$$

Within this conceptual frame Chaitin also proves that Ω is a Martin Löf-Solovay-Chaitin random real. The fact that Ω is inpredictable and incompressible follows from its compact encoding of the halting problem. Because the first n bits of Ω solve the halting problem for all programs of n bits or fewer, they constitute an "axiom" sufficient to prove the incompressibility of all incompressible integers of n bits or fewer. If we consider the axioms of a formal theory to be encoded as a single finite bit string and the rules of inference to be an algorithm for enumerating the theorems given the axioms, by an n-bit theory we can indicate the set of theorems deduced from an n-bit axiom. Remembering that the information content of knowing the first n bits of Ω is $\geq n-c$ and that the information content of knowing any n bits of Ω is $\geq n-c$, we are, finally, able to prove that if a theory has H (Axiom) $< n$, then it can yield at most $n + c$ (scattered) bits of Ω. (A theory yields a bit of Ω, when it enables us to determine its position and its 0/1 value). What we have just said may be summarized from another point of view saying that as we precisely and consistently specify the methods of reasoning permitted, we determine an upper bound to the complexity of our results. In other words we cannot prove the consistency of a particular system by methods of proof which are restricted enough to be represented within the system itself.

In this sense, in Chaitin's conception, we have to step outside the methods of reasoning accepted by Hilbert: in particular we have to extend Hilbert's finitary standpoint by admitting the necessary utilization of certain abstract concepts in addition to the merely combinatorial concepts referring to symbols. For this purpose we must count as abstract those concepts that do not comprise properties or relations of concrete objects (the inspectable evidence), but which are concerned with thought-constructions (connected to the choice of particular measures, the invention of new methods, the identification of the sense of symbols, etc.)

4. As it is quite evident, the real problem, in this context, is how to characterize, in an operational manner, the process of identification of these abstract concepts with reference to the inner articulation of higher-level languages. In this regard a short and necessarily incomplete exemplification will turn out useful.

Let F be a (finite) field of events and let $e_1 \ldots e_n$ be its atoms. As is well known the classical probability's entropy is nothing else that the expected change linked to a specific distance function expressed in a logarithmic scale with reference to the field of events. In Shannon's case the atomic events are treated as anonymous points in an abstract set. But we can assume (Carsetti, 1981; Gaifman, 1982) that the atomic events are of the form $Z = z_i$ where Z is, for instance, a r.v. and the z_i are its possible values whose number and exact identification may depend on time, on changes in the actual world, on changes induced in the composition of the actual world by the actions of the observer and on the improvement of better measurements.

To reflect this situation we should modify continuously our distance functions between probabilities by taking into account varying distances between atomic events. However, once a certain level of complexity is reached, it is impossible to consider in advance the regularities and the internal evolution that some phenomenon might exhibit. They can be viewed only from the point of view of some higher-level language, a language that, as a matter of fact, may in turn be helpful to define new regularities. Information content is not only a property of the information source, but also depends on the ability of the observer to construct sequential tests, to find adequate distance functions and to create new languages.

In this context the crucial point, as we have seen, is that to introduce new axioms with reference to the information-theoretic limits imposed by incompleteness theorem implies to resort to abstract concepts that can be considered as "evidence" inspected not by the senses, but by the intellectuals tools of the human mind.

So far, from an informational and algorithmic point of view, it has turned out very difficult to outline, in a consistent manner, the internal genesis and the structure of the aforesaid abstract concepts that live in the logical space of higher-level languages. However recent developments in the field of research of formal semantics offer some interesting and helpful insight in this regard.

In fact, contemporary denotational semantics (Scott, 1982) and hyperintensional logic (Carsetti, 1981) show how it is possible that some thought-construction can exist as data for other thought-constructions in order to allow the modeling of powerful intellectual tools capable of individuating new aspects of the world-model. In a confirmational context an hypothetical thought-construction will result confirmed and will exist as "object", according to Scott's suggestions, if it will produce maps which reveal themselves as maps monotonically increasing toward the truth. This fact allows us to apply the notions of approximation process, distance function, information content etc. as used for data spaces also in the case of the conceptual structures. So we are able to consistently form compositions of thought-constructions by means of functor application and we can individuate, in an abstract but definite manner, hierarchically links between concepts. In this sense we can finally realize that it is only through a precise process of approximation that the regularities that live in the deep real processes can reveal themselves within a confirmational and "coupled" context, progressively showing their semantic information content. So the new distance functions defined with reference to higher-level languages by means of the approximation process and by means of a continuously renewed identification of fix-points, can be exploited in order to identify new pattern of data and to "guide" the process of progressive revelation of these data. A process that, in turn, in a confirmational context, may act selectively on the development of the intellectual structures, constraining, in this way, from the outside, the path of the inner conceptual creativity.

II University of Rome (Italy)

REFERENCES

Carsetti, A. (1981), "Semantica dei mondi possibili ed operatori iperintensionali". In Atti del Convegno nazionale di Logica, edited by S. Bernini. Napoli: Bibliopolis.

Carsetti, A. (1982), "Semantica denotazionale e sistemi autopoietici". La Nuova Critica 64: 51-91.

Chaitin, G. (1977), "Algorithmic entropy of sets". Comp. and Math. with applications 2: 235-45.

Chaitin, G. (1987), Algorithmic Information Theory. Cambridge: Cambridge University Press.

Gaifman, H. and Snir, M. (1982), "Probabilities over rich languages, testing and randomness". J. Symb. Logic 47: 495-548

Humphreys, P.W. (1977), "Randomness, Independence and Hypotheses". Synthèse 36: 415-26.

Kolmogoroff, A.N. (1956), "Logical basis for information theory and probability". IEEE Trans. Inf. Th. IT-14: 662-664.

Martin Löf, P. (1966), "The definition of random sequences". Inf. and Control 9: 602-619.

Schnorr, C.P. (1971), "A unified approach to the definition of random sequences". Math. Syst. Theory 5: 246-258.

Schnorr, C.P. (1974), Unpublished manuscript.

Scott, D. (1982), "Domains for Denotational Semantics". In Automata, Languages and Programming, edited by Nielsen M. and Schmidt, E.M.: Berlin: Springer.

Solovay, R.M. (1975), Unpublished manuscript.

PART 3

PROBABILITY IN THE NATURAL SCIENCES

Jaques Ricard

PROBABILITY, ORGANIZATION AND EVOLUTION IN BIOCHEMISTRY

Most scientists, if invited to speak to a congress of philosophy of science, vould no doubt be inclined to apologize for their lack of philosophical background. I am not going to do this however, for I am convinced by what the history of logic and epistemology have shown us -that the professional philosopher and the scientist are both obliged to contemplate the general ideas underlying the set of hypotheses, deductions and experiments that constitute the essence of scientific activity. As stressed by Monod, "la modestie sied au savant mais pas aux idées qui l'habitent".[1] Since most biologists believe today that the behaviour of living systems is the expression, at a high level of complexity, of physical laws, one is faced with an apparent paradox that is formidable in both its nature and its implications. If one transposes the laws of physical chemistry to living world, one has to recognize that even the simplest biological processes, such as the elementary steps of metabolism, appear as a collection of highly improbable events. The simultaneous occurrence of all these elementary processes appears as an extremely unlikely event whose probability is close to zero.

Many biologists in the past were not aware of the existence of this paradox. Others have taken advantage of it to favor spiritualist theses,[2] or believe that special laws govern the living world. Some pretend that the origin of life was an improbable event and proclaim that the ancient alliance between nature and mankind has been broken.[3]

The aim of present contribution is basically twofold: first, to try to understand how living systems have progressively devised a molecular machinery that allows them to perform the highly improbable actions involved in metabolism; and second, to

E. Agazzi (ed.), Probability in the Sciences, 175–182.

show that the rather low probability of the occurrence of a biochemical process may give the corresponding organism and unexpected diversity and complexity in its behavior. Of the natural sciences, biology is the one set in an evolutionary perspective. During the evolution of living organisms, molecular devices that represent the very basis of life have evolved. Reflection on the evolution of the behavior of macromolecules during evolution is essentially conjectural, and is certainly located upstream with respect to the criterion of demarcation between metaphysics and science espoused by K. R. Popper, for this kind of analysis cannot be submitted to attempts at falsification. These conjectures however are undoubtly of interest, and therefore deserve consideration here.

Let us consider the reaction of transfer of a group X from molecule A to molecule B:

$$AC + B \xrightarrow{[A\text{--}X\text{--}B]\neq} A + XB$$

This reaction implies that an energy barrier between the initial and the final state is overcome, and pertains to the so-called transition state. This transition state, [A--X--B]≠, is an unstable molecule located along the reaction coordinate midway between the initial and the final states. In order to overcome this energy barrier, the molecules of AX and B must collide with sufficient energy, and must be positioned in the collision complex so as to allow the transfer of X from A to B. The rate constant, k, is directly related to the probability of reaching the transition state. The highest energy is required to overcome the barrier, and the lowest to maintain the reaction rate. Under the usual conditions of temperature, pressure and pH that occur in living cells, the former energy is so high that only a very small number of collisions between reactants are productive and lead to reaction products.

Enzymes that are ubiquitous in living cells considerably accelerate chemical reaction rates. Biological catalysis requires the binding of reactants on the active site of an enzyme. The number of reaction steps involved in the enzyme reaction is thus increased relative to that of the corresponding uncatalyzed reaction, and the height of the energy barrier is thus decreased. Moreover, the binding energy of the reactants to the enzyme is used to decrease the entropy of the system, that is, to position these reactants with respect to one other. Therefore regions of reactant molecules that do not directly participate in the chemical reaction process are involved in the correct positioning of these molecules. The binding energy of reactants to the enzyme is thus indirectly used to accelerate chemical reactions. The importance of positioning, or entropy, effects in the control of biochemical reactions is thus considerable and may increase the reaction velocities by a factor of about 10^5.

These entropy effects were certainly at work in the polymerization reactions of amino-acids and nucleotides that occurred on clay particles in the oceans of primitive earth, some four billions years ago. There is little doubt that the magnitude of entropy effects in a protein molecule is defined by the number of atoms of the polypeptide backbone. Thus proteins will behave as enzymes only if they have a minimum molecular weight of about 20 000 or 30 000. It is obvious that this is not fortuitous and that the precision of reactant positioning on the active site of an enzyme molecule depends on the length and on the flexibility of the polypeptide chain.

The first principle of thermodynamics implies the existence of a complementarity between the structure of the enzyme molecule and the transition state of the reaction, the latter being midway between the substrates and the products. The history of the discovery of this complementarity is illustrative of an epistemological process that has led to a major discovery. As early as 1935 Haldane, and then Pauling, proposed the idea that since an enzyme reaction is more or less reversible, the enzyme molecule must be more or less complementary to the transition state whose structure is midway between that of the reactants in the initial and the final states. This intuition was confirmed theoretically about ten years ago by thermodynamic considerations and, experimentally, by the synthesis of transition state analogs that tightly bind to the enzyme and act as strong inhibitors of the enzyme reaction. The validity of this concept has been confirmed recently by the discovery of abzymes i.e. antibodies endowed with enzyme activity. Upon injecting tc rabbits with transition state analogs of certain chemical reactions, one may increase the number of antibodies that catalyze these reactions. The concept of abzyme represents the direct proof that enzyme activity requires some form of complementarity between the enzyme molecule and the transition state analog.

When an enzyme is in the presence of its substrate and other molecules that resemble that substrate, the enzyme may select a 'wrong' molecule and thereby make an error. This error number is small, and quite compatible with the precision of most of the enzyme-catalyzed reactions. This precision however is not sufficient for a number of reactions that are involved in the processing of genetic information, i.e. for reactions that take part in the expression of genes into proteins. In order to be compatible with life and to keep the frequency of mutations at a low level, these reactions must occur with a very low probability of error. The high fidelity of the molecular mechanisms of genetic expression is reached through the ability of some enzymes, involved in the synthesis of nucleic acids and proteins, to hydrolyze a polymer that has been incorrectly synthesized. This is the phenomenon of 'proofreading' that greatly enhances the fidelity of gene expression.

It is indeed not sufficient that the individual biochemical reactions of metabolism simply occur; in order to constitute a metabolic sequence they must occur altogether. In such a sequence the product of an enzyme reaction is the substrate of the next reaction. If the enzymes of the same metabolic pathway were all physically distinct, the viscosity of the cell milieu would slow down the diffusion of any reaction intermediate with respect to the active site of one enzyme and the corresponding site of the next one. The metabolic chain would then respond very slowly to perturbations of the concentration of reaction intermediates. It is evident that the response should be even slower for large cells. In the case of bacteria the cells are small and the enzymes belonging to the same pathway are usually physically distinct, but owing to the small volume of the cells, the diffusional resistances to the transfer of a reaction intermediate from one site to another are not too high. With the large cells of higher organisms, one may observe the existence of molecular structures that allow to increase the probability of occurrence of a metabolic chain.

Some biochemists such as Albery and Knowles[4] believe that the selection pressure' exerted on living organisms over a period of four billions years has resulted in the increase in the catalytic performances of the enzymes. This is probably not always true, for living systems may derive some functional advantages from rather slow reaction rates, and therefore from low probabilities of occurrence of reactions. This is the case, for instance, when slow conformations changes of enzymes occur far from thermodynamic equilibrium conditions. These confirmation changes may generate hysteresis loops of the reaction rate when it is plotted as a function of the concentration of a reactant. This means that the reaction rate is different depending on whether a ligand concentration is reached after an increase or a decrease of the concentration of that ligand. A similar situation is also obtained if the diffusion rate is retarded by the viscosity of the cell milieu. The existence of hysteresis loops certainly represents an interesting phenomenon because it implies that the enzyme system may 'perceive' not only the concentration of a ligand, but also the direction of a concentration change. When the probability of occurrence of a reaction decreases, the enzyme system may acquire a wealth of different behaviors that is at first sight unexpected. Owing to hysteresis phenomena, event the most elementary organisms may 'perceive' chemical signals from the external milieu and react accordingly.

One may therefore speculate that during evolution, living organisms, on one hand, had settled molecular devices that increase the probability of occurrence of reactions, and that, on the other hand, they have taken advantage of the low probability of the occurrence of these reactions to increase the wealth of their kinetic behavior. Elementary sensing phenomena may be viewed as a consequence of these low probabilities of the occurrence of reactions.

It is certainly of interest to try to situate the ideas that have been developed above in the general frame of the epistemology of contemporary biology. As Crombie[5] and Popper[6] have already outlined, any scientific theory is based on a corpus of ideas that may be considered as preconceived since they are anterior to experiments and may be submitted to the usual criteria of falsification. The appearance of any of these preconceived ideas in the history of science corresponds to a paradigmatic change.[7] Although there exists a certain agreement about the nature of preconceived ideas that serve as the foundations of physical theories, this agreement does not yet exist in biology.

If living systems are considered as complex networks of highly controlled chemical reactions, there exists, as outlined by Harrisson,[8] three types of preconceived ideas which serve as the basis of modern biochemistry. The first of these ideas is structural in nature, the second is based on equilibrium thermodynamics and the third on kinetics and non-equilibrium thermodynamics.

The ideas of the structural type typically correspond to what is now called molecular biology, which is an attempt to explain the macroscopic properties of living systems by reference to the microscopic properties of two types of macromolecules, namely proteins and nucleic acids. As outlined by Prigogine and Stenbers[9] this way of thinking aims at making the theory of the development of living organisms as economic as possible, since one postulates, implicitly or explicitly, that the macroscopie structure of living systems is totally defined by the structure of the genes. This type of preconceived idea typically pertains to what Popper has called *methodological essentialism*[10] which aims at explaining the essence of complex biological phenomena through a number of invariants, or ideas in the Platonic sense of the word. These invariants or ideas are indeed the informational macromolecules. This is precisely what Monod has outlined in his book *Chance and Necessity*[11]. Although this type of approach has had considerable operational success during the last thirty-five years, it has not given, and probably cannot give, a full account of the sphere of biological problems.

The ideas of the structural type are usually associated with the use of equilibrium thermodynamics based on the principle of the minimization of energy. This is understandable in the molecular structural perspective, where individual macromolecules are studied in isolation, precisely under conditions of thermodynamic equilibrium.

Although they are not widely used in biochemistry, because most biologists dislike them, the ideas based on kinetics and non-equilibrium thermodynamics certainly represent an ongoing scientific revolution or paradigm change, in the sense suggested by T. S. Kuhn.[12] In line with these ideas, the nature of the biological macromolecules, which is indeed important, is no longer considered essential. It is the dynamics of chemical networks which is the main focus of interest. The shape and the structure of biological objects and their spatio-temporal organization are considered as the result of a multitude of coupled chemical reactions whose global dynamics represents the main point to be understood. The ideas that have been presented in this contribution clearly belong to this type of approach, which opens new vistas on the role of probability in living systems. These ideas are clearly in line with what Popper has called *methodological nominalism*, which first considers phenomena and their evolution, rather than their essence.

Last but not least, it seems obvious that these different types of preconceived ideas are all required to give a better insight of the nature of life.

Centre de Biochimie et de Biologie Moléculaire du CNRS, Marseille

NOTES

1 Monod, J. (1970)

2 Grasse, P. (1973)

3 Monod, J. (1970)

4 Albery, R.A. and Knowles, J.R. (1976)

5 Crombie. A.C. (1959)

6 Popper, K.R. (1978)

7 Kuhn, T. (1972)

8 Harrison, L.G. (1987)

9 Prigogine, I. and Stengers, I. (1979)

10 Popper, K.R. (1945)

11 Monod, J. (1970)
12 Kuhn, T. (1972)

BIBLIOGRAPHY

Albery, R.A. and Knowles, J.R. (1976), Biochemistery 15: 5631-5640
Crombie, A.C. (1959), Medieval and Early Modern Science. New York: Doubleday.
Grassé, P.P. (1973), L'évolution créatrice du vivant. Paris: Albin Michel.
Harrison, L.G. (1987), Journal of Theoretical Biology 125: 369-384.
Kuhn, T.S. (1972), La structure des révolutions scientifiques. Paris: Flammarion.
Monod, J. (1970), Le hasard et la nécessité. Paris: Le Seuil.
Popper, K.R.:
(1945), The open Society and its Enemies. London: Routledge.
(1978), La logique de la découverte scientifique. Paris.
Prigogine, I. and Stengers I. (1979), La nouvelle Alliance. Paris:Gallimard.

O. Costa de Beauregard

RELATIVITY AND PROBABILITY, CLASSICAL AND QUANTAL

A 'manifestly relativistic' presentation of Laplace's algebra of conditional probabilities is proposed, and its 'correspondence' with Dirac's algebra of quantal transition amplitudes is displayed. The algebraic reversibility of these is classically tantamount to time reversal, or 'T-invariance', and quantally to 'CPT-invariance'. This is closely related to the *de jure* reversibility of the $\rightleftarrows$ negentropy information transition, although *de facto* the upper arrow prevails aver the lower one (Second Law).

1. INTRODUCTION

How is it conceivable that the paradigm of an extended space-time geometry, where occurrences are displayed all at once (which of course does not mean "at the same time") is compatible with the very ideas of chance and becoming? There is in this a technical problem and a philosophical problem, both of which will be addressed .

The possibility of a 'manifestly covariant' probability scheme is afforded by the Feynman (1949) graphs scheme, where the Born (1926) and Jordan (1926) algebra of wavelike transition amplitudes is cast in a relativistic form. Dirac (1930) expressed the Born and Jordan algebra in the form of a 'bra' and 'ket' symbolic calculus which ran, so to speak, almost by itself.

Recently, while thinking about the famous Einstein-Podolsky-Rosen (EPR, 1935) correlations, it occurred to me (1986, 1987) that Laplace's (1774) algebra of conditional probabilities can be presented in a form exactly paralleling the Born-Jordan-Dirac-Feynman algebra. Thus a precise 'correspondence' (in Bohr's sense) can be exhibited between the two algebras, the

E. Agazzi (ed.), Probability in the Sciences, 183–202.

classical one adding partial, and multiplying independent probabilities; and the quantal, wavelike one, doing the same with amplitudes. Let it be emphasized at the very start that the two epithets 'conditional' and 'transitional' are synonymous, and thus exchangeable as suits the context.

Any algebra is by definition timeless, and so no temporal relation is essentially implied in the probability concept.

Algebraically speaking any transition is between two 'representations' of a system (to borrow a convenient expression from the quantal paradigm). In physics, of course, time very often enters the picture; then, one is led to consider the 'preparation representation' and the 'measurement representation' of an 'evolving system', together with (either) the classical 'transition probability', or the quantal 'transition amplitude' from the one to the other. This is equivalent to a 'conditional probability', or 'amplitude', which can be thought of either predictively or retrodictively. So a temporal aspect, when relevant, dramatizes the picture, somewhat like the Greek drama dramatized the metaphysics of Fate.

The formal parallelism, or correspondence, between the classical Laplace and the quantal Dirac schemes breaks down completely at the level of interpretation, because of the basic formula

$$(A|B) = |\langle A|B\rangle|^2 \tag{1}$$

expressing the quantal transition or conditional probability $(A|B)$ in terms of the (complex) transition or conditional amplitude $\langle A|B\rangle$. The off-diagonal terms on the right hand side (which, physically speaking, are interference or beating style terms) entail (as we shall see in more detail) the often 'paradoxical' aspects of (algebraically speaking) 'nonseparability' or (geometrically speaking) 'nonlocality'. Of course, by using an 'adapted representation', one hides these terms, in a way somewhat similar to that in which, by looking in one of three privileged directions, one sees a parallelepiped as a rectangle. But this affords no more than a deceptive semblance of the classical, Laplacean way of handling occurrences. As we shall see in more detail, while the classical summations over mutually exclusive possibilities could be thought of as being over 'real hidden states', quantal summations cannot, and are thus said to be over 'virtual states'.

2. LAPLACE'S 1774 ALGEBRA REVISITED

Expanding upon Bayes's discussion of conditional probability, Laplace wrote a series of papers, the first of which is the famous one of 1774 entitled 'Memoir on the probability of causes by the events'. There, he introduces the concept of what I will call, and denote an intrinsic, reversible conditional probability

$$(2) \qquad (A|B) = (B|A)$$

the probability 'of A if B' or 'of B if A'. Later, however, he discarded this concept in favor of the two converse, usually unequal, conditional probabilities which I denote $|A|B)$, 'of A if B' and $|B|A = (A|B|$, 'of B if A', and which have been used ever since (see Jaynes 1983, formulas A6 and A7 p. 216). These are related to the (essentially symmetric) joint probability $|A)\cdot(B|$ 'of A and B' according to the formula

$$(3) \qquad A)\cdot(B| \equiv |B)\cdot(A| = |A|B)(B| = |A)(A|B|,$$

which reads 'the joint probability of A and B equals the conditional probability of A if B, times the prior probability of B', or the converse. The dot I insert in $|A)\cdot(B|$ is by analogy with a scalar product of vectors.

It turns out that it is very useful to reintroduce Laplace's initial $(A|B)$ concept, and thus to rewrite equations (3) in the form

$$(4) \qquad |A)\cdot(B| \equiv |B)\cdot(A| = |A)(A|B)(B| = |B)(B|A)(A|.$$

Obviously, the two extrinsic conditional probabilities $|A|B)$ and $|B|A)$ are related to the intrinsic conditional probability (2) according to

$$(5) \qquad |A|B) = |A)(A|B)(B|, \quad |B|A) = |B)(B|A);$$

the reason I call them "extrinsic" is that they contain the prior probabilities $|A) \equiv (A|$ or $(B| \equiv |B)$.

Setting B = A we rewrite formulas (4) and (5) as

(6) $|A)\cdot(A| = |A)(A| = |A)(A|A)(A|$,

thereby expressing three things: the joint probability of A and A is idempotent; the intrinsic conditional probability of A if A is unity, i.e. $(A|A) = 1$; and the extrinsic conditional probability of A if A equals the prior probability of A, $|A|A) = |A)$. Thus, introducing complete sets of mutually exclusive occurrences A, ..., and (borrowing a term from quantum mechanics) calling them 'representations of a system', orthonormalization is possible in the form

(7) $$(A|A') = \delta(A,A').$$

No temporal element has been implied up to now. For example, $|A)\cdot(B|$ can denote the joint probability that a U.S. citizen has the height A and the weight B; this is a number, devoid of any idea of succession. If the sampling is refined by considering mutually exclusive categories, such as men and women, the prior probabilities $|A)$ and $|B)$ of A and B in these categories enter the picture, and we end up with formula (4), which is preferable to formulas (3). Thus we can speak of the 'height representation' and the 'weight representation' of a U.S. citizen, and of the transition probability connecting these as being 'naked' $(A|B)$, or 'dressed' $|A)(A|B)(B|$.

'Lawlike reversibility versus factlike irreversibility' (to borrow Mehlberg's (1961) wording) has been discussed by Laplace and, later, by Boltzmann (1898) in essentially equivalent terms, as was noted by Van der Waals (1911). It may be expressed thus:

(8) $|A|B) \neq |B|A)$ *iff* $|A) \neq |B)$.

For example, basket ball players are usually tall and light; therefore, considering high values of height A and weight B, the two extrinsic conditional probabilities $|A|B)$ and $|B|A)$ are far from being equal for this class. This is a logical sort of irreversibility, with no time sequence implied.

Of course time enters the picture if $|A)$ denotes the 'preparation representation' and $|B)$, the 'measurement representation' of a system. In such cases both Laplace and Boltzmann assumed maximal irreversibility, in the form 'all prior probabilities (B| of the final states are equal among themselves', so that there is no need to mention them. This is an extremely radical irreversibility statement which (although quite consonant with the phenomenological claims of Carnot and Clausius) breaks the mathematical harmony - and, even worse, alters the very meaning of the probability concept, as we shall see later.

It is often useful to express a conditional or transition probability (A|C) with an intermediate summation

$$(A|C) = \sum(A|B)(B|C), \tag{9}$$

the word "intermediate" being understood algebraically. For example C can denote the waist measurement of a U.S. citizen, while A and B retain their previous meanings. Formula (9) is known as the generating formula of Markov chains. If a temporal element is implied, it follows from the Laplacean symmetry (2) that the chain can zigzag arbitrarily throughout either space-time or the momentum-energy space (if such a picture is preferred), completely disregarding the macroscopic time or energy arrow. In that case, the word "intermediate" assumes a topological meaning.

As a first example consider the joint probability |A)·(C|, or the conditional probability (A|B), that, in a particular U.S. National Park, we will find at place A the male and at place C the female of a couple of bears. Coupling means interaction. Thus the two bears can meet at some 'real hidden place B' and a summation over B is needed. Logically speaking it makes absolutely no difference whether the AC vector is spacelike, with the meeting place B before, or after, A and C; or whether it is timelike, with, say, B after A and before C. Considered as valid in all three cases, the formula (9) is topologically invariant with respect to V, Λ, or C shapes of the ABC zigzag; and so is formula (4).

As a second example consider two Maxwellian or Boltzmannian colliding molecules. The unnormalized collision probability |A)·(C| is the product of three independent probabilities: the mutual cross section (A|C), and the two initial occupation numbers |A) and (C| or, more generally, their estimated values). This is for prediction.

Retrodictively the same formula holds, with $|A)$ and $(C|$, then denoting the final occupation numbers.

Formulas (4) and (9) also hold for a C shaped ABC zigzag. As previously mentioned $(A|C)$ then denotes the 'naked transition probability', and $|A)\cdot(C|$ the 'dressed transition probability, $|A)$ the initial occupation number of the initial state and $(C|$ the final occupation number of the final state. It so happened that neither Boltzmann, nor Laplace in similar cases, multiplied $(A|C)$ by $(C|$, but both multiplied it by $|A)$. This, as previously suggested, was their (common) way of expressing their belief in 'maximal physical irreversibility'. But this was 'intrinsically illogical' for the following reason: multiplication by $|A)$ does imply 'statistical indistinguishability'; thus, there are $(C|$ ways in which the transiting molecule can reach the final state; therefore, multiplication by $(C|$ is imperative, and is a corollary to multiplication by $|A)$.

Of course, as used in this fashion, formula (4) is none other than the formula prescribed by the quantal statistics - with $|A)$, $|C) = 0,1,2...,n$ for bosons, and $|A)$, $|C) = 0,1$ for fermions. The implication, then, is that the internal consistency of (both) quantal statistics is greater than that of classical statistics - and that the quantal statistics entail the past-future symmetry set aside by Laplace and Boltzmann.

So, in the case of colliding molecules, formulas (4) and (9) have topological invariance with respect to V, Λ or C shapes of an ABC zigzag; the intermediate summation is over 'real hidden states'; in the case of spherical molecules these can be labeled by the line of centers when the molecules touch each other.

A complete Markov chain is expressed as

$$|A)\cdot(L| = \sum\sum \ldots |A)(A|B)(B \ldots K)(K|L)(L|. \tag{10}$$

The Bayesian approach to probabilities has it that the end prior probabilities $|A)$ and $(L|$ are shorthand notations for conditional probabilities $(E|A)$ and $(L|E')$ connecting the system with the environment. Thus we are left with essentially two concepts: the joint probability $|A)\cdot(B|$, and the intrinsic conditional probability $(A|B)$.

3. THE 1926 BORN AND JORDAN ALGEBRA REVISITED

'Corresponding' to Laplace's symmetry assumption (2) we have the Hermitian symmetry

(11) $$\langle A|B\rangle = \langle B|A\rangle^*$$

assumed for transition amplitudes, which are also conditional amplitudes. Going from the real to the complex field not only enlarges the paradigm of probabilities, but, as we shall see, also deepens it.

The joint amplitude of two occurrences A and B is expressed as

(12) $$|A\rangle\cdot\langle B| = |A\rangle\langle A|B\rangle\langle B|,$$

with $|A\rangle$ and $\langle B|$ (The Dirac (1930) ket and bra vectors) denoting prior amplitudes, the absolute squares of which are occupation numbers (or, more generally, the estimated values of these). Setting B = A we get

(13) $$|A\rangle\cdot\langle A| = |A\rangle\langle A| = |A\rangle\langle A|A\rangle\langle A|,$$

showing that $|A\rangle\langle A|$ is a projector, and that $\langle A|A\rangle = 1$; thus, 'orthonormalization' of a 'representation of a system' is allowed, according to

(14) $$\langle A|A'\rangle = \delta(A,A').$$

'Corresponding'to the generating formula (9) of Markov chains we have that one of the Landé (1965) chains

(15) $$\langle A|C\rangle = \sum\langle A|B\rangle\langle B|C\rangle.$$

Whenever a space-time, or a momentum-energy connotation is attached to the occurrences being considered, the Landé chain can, due to the symmetry property (11), zigzag arbitrarily throughout either space-time or the momentum-energy space, disregarding the macroscopic time or energy arrow. Thus formula (15) has topological invariance vis-à-vis V, Λ or C shapes of an ABC zigzag.

An important generalization of the Landé chain is the Feynman (1949) graph, a concatenation, so to speak, where more than two links $\langle A|B\rangle$ can be attached to a vertex A. Topological invariance is a well known property of Feynman graphs.

The end prior amplitudes $|A\rangle$ or $\langle L|$ of a Landé chain

(16) $|A\rangle\cdot\langle L| = \sum\sum ... |A\rangle\langle A|B\rangle\langle B...K\rangle\langle K|L\rangle\langle L|$.

or a Feynman graph are shorthand notations for conditional amplitudes $\langle E|A\rangle$ or $\langle L|E'\rangle$ linking the system to the environment. Thus Dirac (1930) and Landé (1965) write state vectors $\psi_a(x)$ and $\Phi_a(k)$ in the form $\langle a|x\rangle$ and $\langle a|k\rangle$, in the spacetime and the momentum-energy pictures, respectively.

The exact formal parallelism between the 1774 Laplace and the 1926 Born and Jordan algebras, as displayed in the present and preceding section, breaks down completely, however, at the level of interpretation. This is because of the basic formula (1) expressing the quantal transition or conditional probability $(A|B)$ in terms of the transition or conditional amplitude $\langle A|B\rangle$. Let it be recalled that the Born (1926) and Jordan (1926) radically new wavelike probability algebra was adjusted so as to fit the Einstein and de Broglie wave-particle dualism: the addition of partial probabilities had to yield to the addition of partial amplitudes, due to the physical phenomenon of interference or beating.

Because there are off-diagonal terms in the right-hand side of formula (1) (even if formally concealed by use of an adapted representation), the intermediate summations in formulas (15) or (16) can no more be thought of as implying real states; thus they are said to imply 'virtual states'. And this may have very dramatic consequences.

Whenever a space-time or a momentum-energy connotation is attached to the occurrences A,B.,..., the V, Λ and C shapes of an ABC zigzag receive different physical interpretations.

A V shaped ABC zigzag describes what is called an 'Einstein-Podolsky-Rosen (EPR, 1935) correlation' proper between two distant measurements at A and C issuing from a common preparation at B. The so-called 'EPR paradox' (Einstein, 1949, p. 681) consists of the fact that in formula (15) the summation $|B\rangle\langle B|$ over the states inside the source cannot be thought of as implying 'real hidden states'. So, borrowing Miller's and Wheeler's (1983) wording, wee say that it is a 'smoky dragon'.

Accordingly, the states measured as $|A\rangle$ and $\langle C|$ do not preexist in the source. This is why quite a few authors: myself (1953, 1977, 1979, 1983, 1985, 1986, 1987), Stapp (1975), Davidon (1976), Rayski (1979), Rietdijk (1981), Cramer (1980, 1986), Sutherland (1983), Pegg (1980), and possibly others, have introduced the idea of a 'zigzagging arrowless causality'.

A Λ shaped ABC zigzag illustrates a 'reversed EPR correlation' between two distant preparations at A and C merging into a common sink B. Of course, only those paired particles A and C having the right phase relation are absorbed in the sink B; this illustrates 'factlike irreversibility'. To understand what is meant here by 'lawlike reversibility' one must think retrodictively.

For a more detailed discussion of advanced and retarded causality in the context of EPR correlations proper and reversed, I refer to previous publications (1977, 1979, 1987a).

A C shaped ABC zigzag illustrates Miller's and Wheeler's (1983) 'smoky dragon' metaphor. In what 'state' is an 'evolving system' between its preparation as $|A\rangle$ and its measurement as $|C\rangle$? Is it in the retarded state generated by $|A\rangle$ (as classical physics had it)? Or is it in the advanced state converging into $|C\rangle$, as the Hermitian symmetry (11) allows ? It cannot be in both states if there is a transition. And then, due to the symmetry expressed by (11), why should it be in the one rather than in the other?

The truth is that the evolving system is neither in the retarded state issuing from $|A\rangle$, nor in the advanced state passing into $|C\rangle$, because it is actually transiting from $|A\rangle$ to $|C\rangle$. Borrowing an analogy from classical hydrodynamics, we say that it feels symmetrically the 'pressure from the source' $|A\rangle$,and the 'suction into the sink' $|C\rangle$. In other words, it is Miller's and Wheeler's 'smoky dragon', of which nothing definite is known, because its expression is a summation over products $|B\rangle\langle B|$ of virtual states. Only the 'tail' of the dragon held as $|A\rangle$, and its 'mouth' biting as $|C\rangle$, are, so to speak, in our world. The dragon itself lives above, in the $A \otimes C$ Hilbert space. For example, in the well known two slits thought experiment, 'the photon' neither passes through one slit or the other, nor through both. Such images, borrowed from the Laplacean paradigm, have no ready place in the Born.-Jordan-Dirac paradigm.

In the V shaped diagram previously discussed, there is at B, in the source, only a smoky dragon coiled in there, with two mouths biting at A and C; this dramatizes 'the EPR paradox'. In the Λ shaped diagram the dragon, coiled in B, has two tails held at A and C. It is in the nature of dragons to be fantastic.

4. LORENTZ INVARIANCE AND CPT INVARIANCE

That the classical Laplace algebra and the quantal, wavelike, Dirac algebra are both easily endowed with the Lorentz invariance fitting space-time descriptions must be quite obvious by now.

The topological invariance of these two algebras vis à vis the V, Λ and C shapes of an ABC zigzag, in either space-time or the momentum-energy space, leads to a consideration of invariance under the geometrical reversal of all four axes; this can be thought of either as an 'active' or as a 'passive' transformation. The natural guess then is that elementary physical laws are invariant under the geometrical reversal of all four axes.

Classically this can be called 'covariant motion reversal' PT, which (actively) exchanges emissions and absorptions or preparations and measurements, or (passively) exchanges prediction and retrodiction.

Things are more subtle in the quantal paradigm, due to the pairing between particles and antiparticles, and to the Stueckelberg and Feynman expression of their exchange by the reversal of 4-velocities in the x picture, or of energy-momentum in the p = hk picture.

Figures 1a, b, and c show, either in space-time or in the momentum-energy space, the three operations C (particle-antiparticle exchange), PT (covariant motion reversal), and CPT (geometrical, 'strong' space-time reflection $\Pi_{Ⓜ}$, the last expressed by Lüders (1952) as :

$$\Pi_{Ⓜ} = CPT = 1. \quad (17)$$

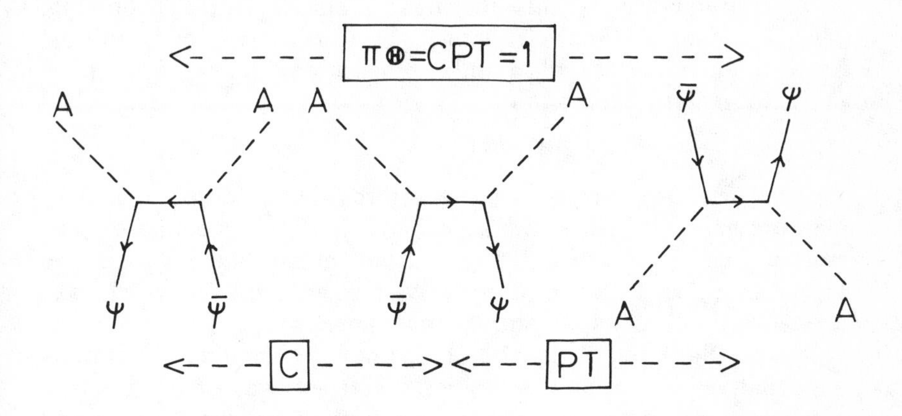

Another symbolization expressing the C, the PT, and the CPT = 1 symmetries is as follows:

(18) C: $\langle A | B \rangle = \langle A | B \rangle^*$,

PT: $\langle A | B \rangle = \langle B | A \rangle$,

CPT: $\langle A | B \rangle = \langle B | A \rangle^*$.

One thus sees that CPT-invariance is built into the very scheme of Hermitian symmetry, and in the second quantization algebra that has been expanded out of it. Extending the relativistic jargon, we can say that C and PT are two 'relative images' of essentially the same operation.

CPT-invariance entails the following 'principle of detailed balance':

(19) $$A + B + \ldots = C + D + \ldots$$

where (beware) a bar means particle on the left hand side and antiparticle on the right side (and vice-versa). This shows that CPT invariance is the faithful quantal-and-relativistic extension of the Loschmidt (1876) T-symmetry. Therefore, *physical reversibility is expressed by the CPT- symmetry, and not by the PT-Symmetry.*

A consequence of this important remark is the following. The Tomonaga (1946)-Schwinger (1948) transition amplitude between a preparation $|\Phi\rangle$and a measurement $|\Psi\rangle$ has the three equivalent expressions:

$$\langle\Psi|U\Phi\rangle=\langle\Psi U|\Phi\rangle=\langle\Psi|U|\Phi\rangle. \tag{20}$$

The first one, projecting the retarded preparation upon the measurement, illustrates the so-called 'state vector collapse'. The second one, projecting the advanced measurement upon the preparation, is a 'retrocollapse'. And the third, symmetric expression, I call 'collapse-and-retrocollapse'.

Fock (1948) and Watanabe (1955) have independently explained that, in quantum mechanics, retarded waves are used in prediction and advanced waves in retrodiction; this is expressed by the first equality (19). Therefore (beware) this Fock-Watanabe symmetry between retarded and advanced waves, and between prediction and retrodiction, is equivalent to the CPT, not the PT symmetry.

Temporal succession, as we have said, provides a dramatization of both the Laplace and the Dirac algebras. In the latter, collapse of the wave function' should not be taken litterally (ipso facto causing much worry with Lorentz invariance). It is merely a computational recipe -and a much less palatable one than the Lorentz-and-CPT invariant transition amplitude concept. Borrowed from the wardrobe of macrophysics, the 'evolving state vector' concept (a cloth loosely draping Wheeler's 'smoky dragon') should be dismissed for being no less noxious to clear thinking than was the late 'luminiferous aether': see Aharonov and Albert (1980, 1981) and myself (1981, 1982).

5. CAUSALITY AS IDENTIFIED WITH CONDITIONAL PROBABILITY OR, BETTER STILL , WITH CONDITIONAL AMPLITUDE

Is there any definition of causality more operational than the following : 'If you do this, then the (predictive) probability that

that will occur is ...'; or : If you find that, then the (retrodictive) probability that this has occurred is ...'? Is there any binding of the subjective and the objective sides of chance that is more subtle or strong than this ?

This is an intrinsically arrowless conception of causality. As previously explained, it affords a topologically invariant rendering of, for example, the classical coupled-bears parable, the colliding molecules problem, and some other cases I have discussed elsewhere (1987). It also affords a topologically invariant rendering of the quantal EPR correlations (proper or reversed), and of Wheeleer's smoky dragon metaphor - all in one stroke. In all these cases, $(A|B) = \sum\langle A|B\rangle\langle B|C\rangle$, or $\langle A|C\rangle. = \sum\langle A|B\rangle\langle B|C\rangle$ are, respectively, the conditional probability or amplitude linking two measurements, two preparations, or a preparation and a measurement (depending on the V. Λ or C shape of the zigzag). Was Laplace perhaps unwittingly anticipating this, when he spoke of the "probabilities of causes" ? In any case, this is the very essence of the S-matrix philosophy.

What we have is a stochastic game, be it classical or quantal, implying questions (the preparations) and answers (the measurements). In between there is, so to speak, the 'wiring of the space-time computer' - that is, in quantum mechanics, the Feynman graph, with Feynman propagators as transition amplitudes $\langle A|B\rangle$.

6. RELATIVITY AND STATISTICAL FREQUENCY

How did the probability concept fit the classical, deterministic paradigm? Two classical stochastic tests were said to be identical when they were the same with respect to all parameters except those assumed to be of negligible influence, the latter being neglected. For example, an aerodynamics engineer repeating a measurement will not worry about the phase of the moon, nor the stock market exchange.

The neglected parameters were allowed to vary, and assumed to do so 'according to the laws of chance'. The probabilistic scheme was thus tested, and either confirmed or 'falsified', the latter, in Laplace's terms, consisting in some 'sufficient reason' having been overlooked. Guessing what this overlooked 'sufficient reason' was amounted to learning something new-eventually something important, for example the violation of the principle of the equipartition of energy and the discovery of the quantum.

In the Bayesian treatment of probabilities this sort of game is called 'hypotheses testing', or 'extracting signal from noise'.

A transposition of these remarks into the field of relativistic probability theories reconciles the use of statistical frequencies with an extended 4-dimensional geometry. Two stochastic tests will be termed identical if they are the same on all parameters up to those considered negligible. In this sense they will be 'translationally deducible from each other'(either in a spacelike or a timelike fashion). In quantum mechanics there is always a set of parameters which are allowed to vary 'according to the laws of chance': those parameters that are conjugate with the ones actually prepared or measured. This is exactly how the Feynman graphs are used at CERN or at Lawrence-Berkeley. Incidentally, two CPT associated Feynman graphs must be thought of as framed pictures, because the environment cannot be CPT reversed, due to two very strong factlike asymmetries: the preponderance of particles over antiparticles, and of retarded over advanced waves.

7. METAPHYSICS OF A RELATIVISTIC PROBABILITY CALCULUS

As there is only one world history, and it makes no sense to think of rewriting it, strictly speaking no frequency interpretation of probability is possible inside a relativistically covariant paradigm. Thus, the only acceptable conception of probability is the so-called 'subjective' one, held by Bernouilli, Bayes, Laplace, and advocated today by Jaynes (1983). This being so, one must take care to note that *information is a two faced concept*: *gain in knowledge* (say, in the transition: *negentropy* → *information* at the reception of a message) and *organizing power* (say, in the transition: *information* → *negenteropy* at the emission of the message). That the one face is so trivial, and the other so much hidden, is a direct consequence of physical irreversibility -and truly the most profound expression of it.

The evaluation of any universal constant in 'practical units' expresses some aspects of our existential-situation-in-the world: that Joule's J 'is' not far from unity means that the First Law belongs to common place physics; but, that Einstein's c 'is' very large means that the relativistic phenomena are largely outside the domain where the meter and the second are appropriate units. So then, what is the meaning of the 'smallness' of Boltzmann's k, or, rather, of the conversion coefficient k Ln 2 between a negentropy N expressed in thermodynamical units, and a piece of information expressed in bits, according to the formula

(20) $$N = (\mathrm{Ln}\ 2)\ k\ I?$$

Letting k go to the limit zero would summarize a theory where gain in knowledge is rigorously costless, and free action utterly impossible - the once fashionable theory of 'epiphenomenal consciousness'. Cybernetics has rediscovered the 'hidden face' of information. It requires consciousness-the-spectator to pay her ticket, at a very low price; and this alone allows consciousness-the-actor to exist, but at exorbitant fees, because the exchange rate applies in the opposite direction.

While the $N \rightarrow I_2$'learning transition' (with de facto if not de jure $N>I_2$) is a generalization of the Second Law, as emphasized by Brillouin (1967), and expresses the familiar 'retarded causality', the $I_1 \rightarrow N$ transition (with then in fact $I_1>N$) expresses 'finality'. This is the very sort of anti-Second-Law phenomenon radically excluded by the irreversibility statements due to Laplace, Boltzmann, Carnot and Clausius - but also the very sort of phenomenon Bergson (1907) and other thinkers deem necessary for understanding 'Creative Evolution'. The intrinsic or lawlike symmetry in the formulas suggests that natural occurrences are, so to speak, symmetrically 'pushed into existence' by efficient causes and 'pulled into existence' by final causes, as Lewis (1930) suggests - somewhat in the way that a hydrodynamical flow is determined jointly by the pressure from the sources and the suction from the sinks.

Accepting this picture does not entail that 'one can kill one's grandfather in his cradle' (as some have insinuated). 'Rewriting history' makes no sense. 'Final cause' does not mean reshaping the past, but does mean shaping it; as Lamarck puts it 'the function creates the organ'.

In other words, 'factlike irreversibility' does not mean suppression, but does mean repression of advanced actions. There are two aspects in an advanced action: 'seeing into the future' (precognition) and 'acting in the past' (psychokinesis; influencing falling dice implies doing so before the outcome is displayed).

These are things that can be, and have been statistically tested in repeatable experiments (Jahn and Dunne, 1987).

The intrinsic reversibility of probabilities implies that ascribing prior probabilities to later occurrences (Laplace's effects') can make sense. The New York Museum of Natural History displays the line of evolution leading to the horse. Given the structure of the 'eohippus' can one predict the structure of the horse? Of course not. But by using Laplace's and Boltzmann's prescription in reverse, one can correctly retrodict the 'primeval molecular soup' - or an equivalent of it. Objectors will then argue that the evolution problem should be placed in context, that systems which are far from equilibrium behave as parasites on the universal negentropy cascade, borrowing negentropy from upstream (Brillouin 1967) and dumping entropy downstream (Nicolis and Prigogine, 1977). All right. Of course, looking at things differently requires reassessing the prior probabilities. But there was nothing illogical in arguing from the eohhippus per se - so little so that, as Boltzmann (1964, pp. 446-448) puts it, "this, by stimulating thought, may advance the understanding of facts".

Others will argue that, after all, the eohippus was not so improbable, as the skeleton of one is displayed in New York.. But this is 'blind statistical reasoning', and *does* imply the finality assumption.

Weighing the prior probabilities of the 'effects' may well be an affair of willing no less than of knowing awareness. Relying as it does on the N → I transition, knowing awareness naturally perceives causality, while using the I → N transition, willing awareness naturally experiences finality. Eccles (1986) argues that will actually consists of biasing the end prior probabilities inside the nervous system, so that it truly is 'internal psychokinesis'. Libet (1985) should also be read in this respect.

Can psychokinesis be observed outside the envelope of our skin ? This is what Wigner (1967, pp. 171-183) implies on the basis of physical symmetry arguments analogous to mine - and is indeed also implied in a statement found in many quantum-mechnanical textbooks: 'Due to the finiteness of Planck's h (and, I would add, also due to the finiteness of Boltzmann's k) there is an unavoidable reaction of the measuring apparatus upon the measured system'. Where should one sever the two ? Between the dial or screen and the eye; somewhere along the optical nerve; or where? So, this seemingly harmless remark entails that *a reaction of the observer* upon the observed system exists in principle.

In the Princeton University School of Engineering there is an 'Anomalies Research Laboratory' where repeatable psychokinetic experiments have been conducted under Dean Robert Jahn, using sophisticated protocols and electronic equipment. His book (1987) makes interesting reading and, in my opinion, may well herald a scientific breakthrough.

Institut Henri Poincaré (France)

BIBLIOGRAPHY

Aharonov, Y. and Albert, D.Z. (1980), 'States and Observables in Relativistic Quantum Mechanics'. Physical Review 21: 3316-3324.

Aharonov, Y. and Albert, D.Z. (1981), 'Can we make sense out of the Measurement Process in Relativistic Quantum Mechanics'. Physical Review 24: 359-370.

Bergson, H. (1907), L'Evolution créatrice. Alcan, Paris.

Boltzmann, L. (1964), Lectures on Gas Theory (transl. by S. Brush). Univ. of California Press.

Born, M. (1926), 'Quantenmechanik des Stossvorgänge'. Zeitschrift für Physik 38: 858-888.

Costa de Beauregard, O.:

(1953), 'Une Réponse à l'Argument Dirigé par Einstein, Podolsky et Rosen Contre l'interprétation bohrienne des Phénomènes Quantiques'. C.R. Académies des sciences 236: 1632-1634.

(1977), 'Time Symmetry and the Einstein-Podolsky-Rosen Paradox'. Il Nuovo Cim. 42B: 41-64.

(1979), 'Time Symmetry and the Einstein-Podolsky-Rosen Paradox, II'. Il Nuovo Cimento 51B, 267-279.

(1981), 'Is the Deterministically Evolving State Vector an Unnecessary Concept?, Lettere al Nuovo Cimento 31: 43-44.

(1982), 'Is the Deterministically Evolving State Vector a Misleading Concept'. Lettere al Nuovo Cimento 36: 39-40.

(1983), 'Lorentz and CPT Invariances and the Einstein-Podolsky-Rosen Correlations'. Physical Review Letters 50: 867-869.

(1985) 'On Some Frequent but Controversial Statements concerning the Einstein-Podolsky-Rosen Correlations'. Foundations of Physics 15: 871-887.

(1986), 'Causality as Identified with Conditional Probability and the Quantal Nonseparability'. Annals of the New York Acad. of Sciences 480: 317-325.

(1987a), 'On the Zigzagging Causality EPR Model: Answer to Vigier and Coworkers and to Sutherland'. Foundations of Physics 17: 775-785.

(1987b), Time, The Physical Magnitude. Reidel: Dordrecht.

Cramer, J.G. (1980), 'Generalized Absorber Theory and the Einstein-Podolsky-Rosen Paradox'. Physical Review 22: 362-376.

Cramer, J.G. (1986), 'The Transactional Interpretation of Quantum Mechanics'. Reviews of Modern Physics 58: 847-887.

Davidon, W.C. (1976), 'Quantum Physics of Single Systems'. Il Nuovo Cimento 36: 34-40.

Dirac, P.A.M. (1930), The Principles of Quantum Mechanics. Clarendon Press: Oxford.

Eccles, J. 'Do Mental Events Cause Neural Events Analogously to the Probability Fields of Quantum Mechanics?'. Procedings of the Royal Society 22: 411-428.

Einstein, A (1949), 'Reply to Criticisms'. In Albert Einstein Philosopher Scientist, edited by Shilpp P.A.. Illinois: The Library of Living Philosophers: pp. 665-688.

Einstein, A., Podolsky B. and Rosen, N. (1935), 'Can Quantum Mechanical Description of Physical Reality be Considered Complete'. Physical Review 48: 777-780.

Feynman, R.P. (1949a), 'The Theory of Positron'. Physical Review 76: 749-759.

Feynman, R.P. (1949b), 'Space-time Approach to quantum Electrodynamics'. Physical Review 76: 769-789.

Fock, V. (1948), 'On the Interpretation of the Wave Function Directed Towards the Past'. Doklady Acad. Nauk SSSR 60: 1157-1159.

Jahn, R., and Dunne, B.J. (1987), Margins of Reality. Harcourt: Brace, Jovanovich.

Jaynes, E.T. (1983), Papers on Probability, Statistics and Statistical Physics, edited by Rosenkranz, R.D.. Dordrecht: Reidel.

Jordan, P. (1926), 'Ueber eine Begründung des Quantummechanik'. Zeitschrift für Physik 40: 809-838.

Lande, A. (1965), New Foundations of Quantum Mechanics. Cambridge: Cambridge University Press.

Laplace, P.,S. (1891), 'Mémoire sur la Probabilité des Causes par les Evénements'. In Oeuvres Complètes, Paris: Gauthier Villars, vol. 8: pp. 27-65.

Lewis, G.N. (1930), 'The Symmetry of Time in Physics'. Science 71: 570-577.

Libet, B. (1985), 'On Conscious Will in Voluntary Action'. Behavioral and Brain Sciences 8: 529-615.

Loschmidt, J. (1876), 'Ueber der Zustand des Warmgleichgewichtes eines Systems von Körpern mit Rücksicht auf die Schwerkraft'. Sitz. Adak. Wiss. in Wien 73: 139-145.

Lüders, G. (1952), 'Zur Bewegungsumkehr in Quantiesierten Feldtheorien'. Zeitschrift für Physik 133: 325-339.

Mehlberg, H. (1961), 'Physical Laws and the Time Arrow'. In Current Issues in the Philosophy of Science, edited by Feigel, H. and Maxwell D.. New York: Holt Reinhart, pp. 105-138.

Miller, W.A., and Wheeler, J.A. (1983), 'Delayed Choice Experiments and Bohr's Elementary Phenomenon'. In Proceedings of the International Symposium on the Foundations of Quantum Mechanics in the Light of New Technology, edited by Kamefuchi S. and alii. Tokyo: Phys. Society, pp. 140-152.

Nicolis, G. and Prigogine, I. (1977), Self-Organization in Non-Equilibrium Systems. New York: Wiley.

Pegg, D.T. (1980), 'Objective Reality, Causality and the Aspect Experiment'. Physics Letter 78A: 233-234.

Rayski, J. (1979), 'Controversial Problems of Measurement within Quantum Mechanics'. Found. Phys. 9: 217-236.

Rietdijk, C.W. (1981), 'Another Proof that the Future can Influence the Present'. Found. Phys. 11: 783-790.

Schwinger, J. (1948), 'Quantum Electrodynamics, I: a Covariant Formulation'. Physical Review 82: 914-927.

Stapp, H.P. (1975), 'Bell's Theorem and World Process'. Il Nuovo Cimento 29B: 270-276.

Sutherland, R.I. (1983), 'Bell's Theorem and Backwards-in-Time Causality'. Intern. J. Theor. Phys. 22: 377-384.

Tomonaga, S.I. (1946), 'On a Relativistically Invariant Formulation of the Quantum Theory of Wave Field'. Prog. Theor. Phys. 1: 27-42.

Waals, J.D. van der, 'Ueber die Erklärung des Naturgesetze auf Statistisch-Mechanischer Grundlage'. Phys. Zeitschrift 12: 547-549.

Watanabe, S. (1955), 'Symmetry of Physical Laws'. Reviews of Modern Physics 27: 26-39.

Wigner, E.P. (1967), Symmetries and Reflections. Cambridge Massachussets: M.I.T. Press.

Massimo Pauri

PROBABILISTIC ONTOLOGY AND SPACE-TIME: UPDATING AN HISTORICAL DEBATE

One reason for reconsidering the traditional debate about the nature of space and time at a Meeting on "probability in the sciences" is simply that present scientific knowledge in microphysics clearly suggests that probability is fundamentally related to the very concepts of space and time.

In fact, there are strong indications or even compelling reasons favouring the combination of the probabilistic ontology of quantum theory with the microscopic (or perhaps ultra-microscopic) structure of space-time, even though up to now we have not reached any definite conclusions in this regard. At present we have a fundamental classical deterministic theory of space-time on the macroscopic level (General Theory of Relativity: *GTR*) and a fundamental Relativist Quantum Theory of fields and particles (*RQFT*) on the microscopic level, the latter apparently relying on a *provisional* model of 'microphysical' space-time. As far as their principles are concerned they are essentially conflicting theories which, despite attempts dating back to 1927 an a number of recent exciting suggestions, resist synthesizing in the form of a consistent Quantum Theory of Gravity. Both of these theories tend to have an unrestricted domain of validity which can exert itself only if they mutually ignore each other; furthermore any semi-classical theory of gravity which tries to reach a compromise between them in a pragmatic way appears to be inconsistent.

It seems worthwhile in this connection to take the opportunity offered by this Meeting to comment briefly on the repercussions that the present situation and its possible evolution may have on the traditional debate between the absolutist and relational views of space-time. Actually, the whole philosophical struggle which has been going on during this century (from

E. Agazzi (ed.), Probability in the Sciences, 203–215.

Poincaré and Russell to Reichenbach, Carnap, Grünbaum and many others) -not to mention the original philosophical rivalry between Newton and Leibniz- has for no apparent reason been confined to the macroscopic level and to classical theories (the *GTR* in particular). Independently of the various expectations from Quantum Gravity, however, the very existence of the currently accepted *RQFT* has, I believe, profoundly changed the complexion of the debate. This is essentially due to the ontological reduction, inherent to *RQFT*, of the concept of micro-object, and the paradoxical relation between the micro-world and the classical macroscopic level of empirical observations that the reduction implies.

Let us consider the basic issue. The historical controversy focuses on the following rival conceptions of space and time. According to the first, space and time are independently existing *entities* endowed with an intrinsic structure, which contain and are independent of material objects and processes (this is the *absolutist* conception). According to the second, space and time are conceptual *abstractions* dependent upon material objects and processes, the *relations* among which they hypostatize (the *relational* conception).

Disregarding historical antecedents, while the first view is obviously to be traced back to Newton, two profoundly different philosophical traditions must be carefully distinguished within the second: an *empiricist* or even *phenomenalist* line which can be credited to Berkeley and Mach, and an *ontological* line originated by Leibniz[1].

Furthermore, as to the historical absolutist view, at least three meanings of the term 'absolute' should be distinguished: 1) an *ontological* meaning to the effect that the *identity* of points and instants is *autonomous* and not *derived* from objects and events. Accordingly, space and time each possess their own *intrinsic metric*, independent of the existence of material devices for measuring length and duration; these devices consequently play the purely *epistemic* role of aiding in the description of this independently existing structure, and do not ontologically *constitute* it[2]; 2) a meaning connected to the *homogeneity* or *uniformity* of space and time and therefore to properties of *invariance* with respect to certain transformations, which properly represent a form of *objectivity* and, finally, 3) the meaning of *dynamical independence* of the structure of space and time from the existence of matter (in a wide sense) and from its evolution; in particular, Newtonian space is indefinitely penetrable ('empty') but also *rigid* in the sense that the distances between *unoccupied* positions are invariable and determine the frame for the causal relations among bodies (and similar concepts hold for time).

Now, while the advent of the *GTR* has largely done away with the third meaning of 'absolute' for the *macroscopic* notion of space-time, the question of the first meaning and the interpretation of the second[3], together constitute the focal point of the present controversy.

According to the most recent and rigorous empiricist point of view, the original (or 'standard') *GTR* historically represents the first concrete scientific realization of the relational view in the following specific sense[2]: a) bodies and light rays first define i.e. *individuate* points and instants by conferring their identity upon them by means of coincidences, thus enabling them to serve as the loci of other bodies and events: b) space and time are *metrically "amorphous"* pending explicit or tacit appeal to the bodies which have first to define their respective metrics. Accordingly, space-time is represented by a manifold which presupposes the Cantorean continuum, a topology and therefore a dimensionality, a differentiable structure but *not* a *metric*. The metric must be conferred to space-time *extrinsically* by means of an extra-geometric attribute. In other words, what validates the specific interpretation of the solution of the Einstein equations as *the metric* tensor of space-time is the *ontological mediation* of other *macroscopic* entities such as e.g. 'practically rigid rods' and 'standard clocks' which assume, at least ideally, a *constitutive* role. It is only after the choice of a particular congruence standard, whose rigidity or isochrony under transport is settled by *stipulation*, that the question of an empirically or factually determinate chrono-geometry can be asked. In this sense the chrono-geometry is said not to be concerned with space-time itself but with *relations* among *bodies* and consequently to support the relationalist thesis. From this point of view, the *mathematical* objectivity, in the second sense of 'absolute' treated above, which remains in the GTR through the 'invariants' of the Riemann tensor, could not warrant the attribution of *physical* objectivity to the chrono-geometry. In fact, it would be possible to consider certain different chrono-geometrical descriptions which are prima facie *incompatible* (e.g. Euclidean/non-Euclidean) as alternative *equivalent* descriptions of the same *factual content* specified by the coincidence behaviour of transported bodies. This would require however the adoption on purely pragmatic grounds of specific different stipulations (or 'coordinative definitions') concerning *congruence standards*. (For example, the Einstein choice would correspond to a maximization of the simplicity of the congruence definition, in the sense that it would be possible to retain the Euclidean description by allowing the congruence definition to become sufficiently complicated.)

It is clear that the empiricist relationalist thesis depends heavily on the practical individuation of a family of 'concordant external metric standards' physically *realizing* the Einstein field. Now, as has been discussed at length in the literature[4], this individuation is not free from ambiguities and possibly from logical circularities which seriously weaken the conceptual attractiveness of the entire point of view. Actually, it requires an open-ended correction procedure involving a whole set of physical laws parasitic on Euclidean geometry and non-relativistic theories; and, what is worse, it implies the *GTR* itself. But this precisely means that the *metric* too must be considered a logical component of the theory on the same footing as the topological structure, and that it must borrow its meaning implicitly from the empirical significance of the theory as a whole. This constitutes yet another blow to the empiricist thesis according to which it should always be possible to isolate a privileged set of properties (e.g. topological and order relations) which are declared *intrinsic* and *objectively existent*, as compared to certain other properties (e.g. congruence and metric) which are said *not* to exist objectively and to be ontologically dependent on the former properties through stipulations or coordinative definitions.

It seems therefore very difficult on these classical grounds alone to sustain the above relationalist interpretation of the

GTR. In addition, just from a relationalist point of view, the fact that in this way the *dimensionality* of space-time happens to play an intrinsic and absolute role instead of being deduced from properties of matter should be considered highly unsatisfactory. Independently of these problems, however, it must be stressed that the empiricist defense of relationalism within the standard *GTR* is strictly correlated to the fact that this theory contains a purely phenomenological description of matter, with the consequence that the measuring devices upon which its empirical foundation rests are not, theoretically, self-sufficient entities. Yet, far from providing logical support for the empiricist relationalist program, as the literature might lead one to believe, this fact appears now to create another basic problem for the program. This is because any residual ontological role for the foundation of the metric which could possibly be attributed to medium-sized macroscopic objects would be undermined by the fact that their description as given by standard *GTR* must be considered only *provisional*. In fact -as present scientific knowledge dictates- the properties of matter *must* be explained in terms of quantum theory (*QT*). Therefore it appears that *QT* comes to the fore by force of the argument itself.

Now, what happens to *QT*? Let us consider first the currently accepted theory. It is clear that the *ontological contingency* of quantum micro-objects makes any relational conception of space-time, in the above mentioned empiricist sense, impossible. In order to formulate the probabilistic propositions of *QT*, space-time has to be presupposed as an independent *background* in which different histories of micro-objects could develop, or better, in which states of four dimensional fields could be defined. Moreover it is obviously impossible to *constitute* points and instants in advance according to the relational strategy of the *GTR* through coincidences determined by intersections of world-lines of *definite identity*: an a priori structure of *unoccupied* points and instants seems therefore essential. In fact, the concept of the space-time continuum of relativistic *QT* is constructed through a microscopic *analogical utilization* of the classical *macro* space-time of the special theory of relativity (*STR*), a fact which is particularly evident from the fundamental notion of 'micro-causality' which lies at the basis of the theory. The mathematical structure of this *continuum* functions like a 'translator' from the symbolic level of *QT* into the macroscopic space-time language of the irreversible traces in which the empirical findings are formulated. Having been introduced into the theory by means of group-theoretical techniques in order to satisfy the relativistic invariance of quantum measurements with respect to *macroscopic* reference frames, this continuum structure is tied to the macroscopic medium-sized objects which asymptotically define the experimental setting but it constitutes an 'absolute' entity in an even *stronger* way than for Newton: besides being absolute in all of the old ways, it *selects* the a priori possible structures of micro-objects, hence of matter, through the 'quantum-representations' of its own symmetries.

In essence, this is nothing but another manifestation of the well known semantic inconsistencies and epistemological paradoxes of *QT*. As is well known, although from a logical-ontological point of view the micro-physics should systematically precede the macro-physics, the currently accepted *QT* can be formulated only after and starting from the macro-physics. But *QT* does not provide a *theoretical* definition of the *threshold* between the micro and macro-world, so the distinction turns out to be only a *generic* one. As a particular consequence, the theory does not guarantee that the macroscopic states of large bodies (and of space-time itself, if eventually quantized) are 'objective' in the same sense as are classical states, i.e. free from quantum interference effects. Actually, *QT* reproduces the classical behaviour of medium-sized macroscopic objects and, in particular, of the measuring apparatuses on which it is empirically founded, only in the *formal* sense of mathematical limiting procedures performed within *isolated* sectors of the quantum formalism and for certain subsets of quantum states. Therefore it provides at most a proof of the *theoretical possibility* that classical physics could be an *approximation* of *QT*, without exploiting the physical conditions of this approximation in the form of a consistent quantum theory of measurement.

Correspondingly, within the *macroscopic* framework the space-time concept of *STR* is considered as a local approximation of the space-time described by *GTR* (which, partially at least, depends on the distribution of bulk matter and radiation), while within the *microscopic* framework the *same* conceptual construct assumes an absolute role for the description of quantum micro-objects (which, from a logical-ontological viewpoint, should be considered the primary constituents of macroscopic matter). For the time being, the medium-sized macroscopic objects which, for the sake of theoretical and systematic unity, should be described in the Euclidean language of classical mechanics as approximations with respect to both sides of the dimensional scales, have a purely plausible existence within *GTR*, and can be described only in a formal way by the microscopic theory. Thus they maintain an ambiguous primitive role for the empirical foundation of both theories, and consequently also for the physical concepts of space and time. For these reasons, although it is apparent that the very existence of *QT* has notably eroded the epistemological basis of the empiricist relationalism associated with the historical formulation of the *GTR*, any definite philosophical conclusion about the dichotomy between the absolutist and the *empiricist* relational views in general, appears to be dependent on the particular interpretation of the semantic inconsistencies of *QT* itself, and is likely to be strongly influenced by the future development of the quantum theory of measurement. At the same time, however, the issue over the absolutist conception and *ontological* relationalism, far from being of purely historical interest, seems still alive and even more sophisticated than ever, because of its connection with the eventual construction of a true *physical* concept of *microscopic* space-time within a consistent theory of quantum gravity.

At this point we should mention an important epistemological transformation undergone by the *GTR* during its historical development. The difficulties of the empiricist interpretation of the standard *GTR* led Einstein himself in his last years to reverse his original point of view towards a 'monistic' self-interpretation of the *GTR*, according to which matter and radiation should be exhaustively explained in terms of the metric field alone. The monistic point of view of the integral 'geometrization' has been fully exploited by Wheeler and his co-workers in their 'Geometrodynamics'[5]. According to this last reincarnation of the visions of Parmenides and Descartes, masses, charges and fields, in principle the whole of classical physics, should be interpreted in terms of the 'curvature' and the 'multi-connectedness' of space-time. It is clear that if the *GTR* could succeed in providing a fundamental description of medium-sized bodies in terms of the metric field, they would be -to quote Grünbaum- literally *ontologically* reduced to being part of the field, and so could no longer relationally confer any geo(*metric*) status on the field which it does not already have. Thus the congruence relations would be *determined* by the theory and the measuring devices would acquire a purely *epistemic* role with respect to the metrical field. Actually the same operational complementarity between 'field' and 'test particle' would be reduced to a purely epistemic distinction within an approximation procedure in an otherwise 'closed' description, Thus in this perspective the metric field would assume a fundamental *ontological* bearing, becoming the unique *autonomous* substance of the physical world, and it could even provide the premises of an eventual *empirical* refutation of the relationalist thesis.

Now, while the geometrodynamical program cannot be clearly realized in a classical framework, it is nevertheless important in that it changes our way of thinking about the metric field, and in that it may open a fully *relational* perspective when coupled with the quantum principle. Roughly speaking, the phenomenalist point of view is based on the assertion that space and time 'per se' cannot be sensibly *perceived* and consequently cannot be *real*, while bodies and events can be associated with more or less stable perceptions from which they derive their right to the status of partaking in *reality*. The fact that in this way a true ontology of the bodies is precluded has its scientific counterpart in that specific empirical devices such as 'pratically rigid rods' and 'standard clocks' play the part of primitive entities. On the other hand, for Leibniz, space and time are pure *order relations* and are not real entities in the ontological sense, since they do not contain in themselves the 'principle' of their reality; according to his vision, however, they should *not* be naively reduced to an hypostatization of relations among ontologically *prior* material entities. Actually, for Leibniz, neither space-time nor bodies are real entities, since both are generated via the 'monadic' representative processes. And the *monads* which are the only real entitites, are devoid of any spatial or temporal constitution. Therefore, according to his ontological relationalism, both the bodies ('materia secunda') and space and time are *secondary* or phenomenal entities which derive their reality from a deeper level of Being of a purely *logical* nature. Admittedly Leibniz, betraying his own metaphysics, tried to implement his vision within a scientific context by founding the concept of relation upon *bodies*, and thus failed to free his ontological relationalism from macro-empirical factors; but, of course, although he was clearly aware of the logical need of a deeper *structure*, he could not do otherwise since he was unaware of the quantum level of reality.

Finally, a word should be said about the present perspectives. The most important attempts to construct a quantum theory of gravity (*QG*) can be divided into two main research programs, the 'covariant' and the 'canonical'. From the present point of view, they can be characterized by the fact that they tend to shape different concepts of *microphysical* space-time corresponding to different *degrees* of *involvement* of probability.

The 'covariant' program involves a number of approaches, from the more standard ones based on the *RQFT* for the spin-2 field in a flat or curved *pre-assigned* classical background space-time, to the so-called *pre-metric* theories and the recent multi-dimensional super-symmetric relativistic 'string' theory. These approaches all share a common feature: whether as a consequence of the approximation technique employed, or upon explicit theorization, they involve some splitting among the various mathematical and physical layers and functions which were unified within the classical concept of the gravitational-inertial field of the *GTR*. This is a consequence of two conflicting needs: that of dealing with the gravitational field as a force-field on the same footing as all other fields; and that of treating it as a quantum variable which should establish, as *a result* of the dynamics, the *time-ordering* of the *causal* relations and the propagation properties of *all* fields. The 'intentio logica' of the more recent approaches is to generate structural elements of space-time dynamically from some unique and massless quantum entity containing a 'fundamental length', and in particular to recover the *macroscopic dimensionality* and the *metric* function within some low energy (i.e. large scale) approximation of the theory. But, as a matter of fact, besides the 'length', they *presuppose* at least locally a dimensional extension of Minkowski space-time or a 'conformal structure' (light-cones) which is essential for defining temporal order and propagation properties of the entities considered as fundamental. These manifold structures, although possibly conditioned by dynamical factors and admittedly rather vague as to their epistemological status, would persist as *absolute* backgrounds.

On the other hand the canonical approach, corresponding to the quantum version of Geometrodynamics, involves a universal quantization of the chrono-geometry. Now the analogue of the coordinate 'q' of ordinary *QT* is a whole 'intrinsic 3-geometry' i.e. a 3-space (which is the real dynamic variable), while the analogues of the momentum 'p' are the magnitudes which control the way in which 3-geometry 'unfolds' into a '4-geometry' i.e. *a* space-time, thus *generating* 'time'. Correspondingly, Heisenberg's inequalities imply an indeterminacy or ontological contingency of space-time itself. Thus it becomes impossible to predict or assign any meaning at all to the idea of a history of the deterministic evolution of space in time. What collapses therefore is the *duration* of space in time, with the consequence that space-time becomes a purely *virtual* entity. Strictly speaking, if the space-time interval becomes an operator, the very relation between 'equivalent observers' becomes indeterministic -a fact that shows to what extent the *GTR* and *QT* conflict, and which raises very difficult problems of interpretation in connection with the *empirical* notions of time and space. This contingency or *virtuality* of space-time in turn calls for a resolution of the puzzles connected to the unrestricted allowability of quantum-coherent superpositions of macroscopically distinguishable states. This is so since an indeterminacy of the space-time structure at the macroscopic level (i.e. an indeterminacy not confined to microscopic or better Planckian scales, 10 ·. exp(-33) cm.) would result in the most striking of the quantum paradoxes, and would be clearly unacceptable with respect to a wide range of philosophical positions.

Unless one resorts to the so-called 'many-worlds' interpretation of *QT*[6] -which I personally cannot share- this provides strong support for the idea that constructing a quantum concept of space-time requires some 'reformation' of *QT* itself (where it should be clear that I do not at all mean the 'hidden variables' ideology). This fact may, after all, represent a welcome state of affairs. Actually, the energies involved in *QG* appear to be beyond direct experimental control for any foreseeable technological future. It seems therefore that the only hope of obtaining a physically meaningful theory will depend on the convergence about this problem of some great unresolved questions of principle. In fact only something like this would confer to the theory the conceptual rigidity necessary to make any low-energy experimental test of it empirically probative. Such great questions could be the unification of the fundamental forces, the mass spectrum of the 'elementary particles', the connection of 'spin' with the topological and metrical structures, evolutionary cosmology and, finally, a nomological explanation of time's *arrow*.

Anyway, should the essence of the canonical approach be empirically confirmed, we would have an *ontological overturn* of Geometrodynamics. This would mean that *all* space-time properties -continuity, dimensionality, differentiable structure and metric- could eventually be viewed as *secondary* and 'probabilistic' manifestations of some deeper quantum level of a *pre-geometric*[7] nature, on which space-time would be parasitic, ontologically no less that epistemologically. In this case a fully relational perspective in the Leibnizian (ontological) sense would emerge in which 'probability' would play a far more fundamental role than ever previously imagined. The historical Leibnizian vision which failed when based upon (macroscopic) *bodies*, could be vindicated by the quantum *substructure* of the world.

Historically, every change in the concept of what it is to be an *object* has induced, sooner or later, a corresponding transformation of the concepts of space and time. It appears that the transformation corresponding to the ontological reduction introduced by the concept of quantum 'micro-object' has not, as yet, been fully exploited.

University of Parma, Physics Department (Italy)

NOTES

1 Of course the Kantian conception of space and time, which could be defined as 'functionalist', is neither absolutist nor relational. I purposely leave it aside in the present discussion : although of paramount importance from the philosophical point of view, exploiting its influence and implications on present scientific thought about space and time would require a very rigorous analysis which is beyond the scope of this paper.

2 In this regard cf. Grünbaum, A. (1971).

3 I am not specifically concerned with a discussion of the relativity of simultaneity. The basic questions connected with the meaning of 'absolute' considered here are naturally tranferred from space and time to space-time.

4 Cf. e.g. Grünbaum, A. (1974) and Friedman, M. (1983).

5 Cf. Wheeler, J.A. (1962).

6 Cf. Smolin, L. (1984).

7 Cf. von Weizsäcker, C.F. (1971); Penrose, R. (1975); Finkelstein, D. and Rodriguez, E. (1986).

BIBLIOGRAPHY

Finkelstein, D. and Rodriguez, E. (1986), 'Quantum Time-Space and Gravity'. In Quantum Concepts in Space and Time, edited by Penrose, R. and Isham, C.J.: Oxford: Clarendon Press.

Friedman, M. (1983), Foundations of Space-Time Theories: Princeton: Princeton University Press.

Grünbaum, A.:

(1971), 'Geometry, Chronometry and Empiricism'. In Minnesota Studies in the Philosophy of Science, vol. III, edited by Feigl, H., and Maxwell, G.: Minneapolis: University of Minnesota Press.

(1974), 'Philosophical Problems of Space and Time'. In Boston Studies in the Philosophy of Science, vol. XII, edited by Cohen, R. S. and Wartofski, M. W.: Dordrecht: D. Reidel.

Penrose, R. (1975), 'Twistor Theory, Its Aims and Achievements'. In Quantum Gravity: an Oxford Symposium, edited by Isham, C.J., Penrose, R. and Sciama, D.W.:Oxford:Clarendon Press.

Smolin, L. (1984), 'On Quantum Gravity and the Many-Worlds Interpretation of Quantum Mechanics'. In Quantum Theory of Gravity, edited by Christensen, S.M.: Bristol: Adam Hilger.

von Weizsäcker, C.F. (1971), 'The Unity of Physics'. In Quantum Theory and Beyond, edited by Bastin, T.: Cambridge: Cambridge University Press.

Wheeler, J. A. (1962), Geometrodynamics. New York: Academic Press.

Alberto Cordero

PROBABILITY AND THE MYSTERY OF QUANTUM MECHANICS

1. INTRODUCTION

Measurement induced transitions, incompatible properties, radical event indeterminateness, and yet well defined probabilistic behavior at all levels; the quantum world looks at first sight very mysterious indeed. I shall suggest, however, that a good proportion of the quantum mystery is man-made, brought on by a host of discredited expectations about science and the world.

2. THE EPR ARGUMENT

Einstein was convinced that behind the probabilistic descriptions of QM [quantum mechanics] lies an objective world made up of fully determined and separable components, governed at all levels by the same local deterministic laws. QM may offer a mysterious picture of the world, but -Einstein thought- the mystery would prove superficial if it were possible to establish that QM cannot be a complete theory of the world. And yet, how could one establish this? Einstein's attempts to dissolve the quantum mystery found a famous expression in the Einstein-Podolsky-Rosen argument, EPR argument for short (Einstein et al. 1935).

The EPR argument purports to establish that QM is an incomplete theory. Here I shall limit myself to an outline of the argument as it applies to a particular case[1].

Suppose two previously isolated protons (symbolized as 1 and 2) interact with the result that they end up in the following "compound" quantum state (zero total spin, singlet-state):

E. Agazzi (ed.), Probability in the Sciences, 217–235.

$$Sx(1+2;0) = Sx(1;+)Sx(2;-) - Sx(1;-)Sx(2;+),$$

where 'x' represents an arbitrarily chosen direction, and 'Sx(N;s)' stands for 'spin component along direction x for system N, value s'; since we are talking about protons, s can only take two values, which are represented as "+" and "-".

Let us consider the EPR argument for the case in which two protons are prepared in this state and then left to spread out over a large region of space. If under these conditions a measurement of Sx is performed on one of the protons, a certain value (say +) is obtained. In addition, according to QM the measurement process brings the system into one of the two states superposed in the singlet state, with the further consequence that the position and spin of the second proton are also determined. What the EPR argument purports to show is that the determinability of this second value constitutes proof that the spin components were not created by the act of measurement but had precise values all along.

The proof proceeds roughly as follows. Even if the two protons were not separate entities immediately before measurement (if, for example, they constituted a material system spread out or otherwise blended in space), their total system would still be endowed with separable parts. But then, given that all interactions are ultimately local (Principle of Locality), the alleged constitution of the second proton and its properties at the time of measurement can only be a function of what was available in the corresponding vicinity immediately before; since nothing happens before, however, it is concluded that properties like the proton's spin must have been fully determined all along. The argument can be easily generalized to conclude that all the classical properties must be thought of as being present before measurement, with the implication that, since the rules of QM are incompatible with the simultaneous ascription of all these properties, then QM can only be an *incomplete* description of physical reality.

According to EPR, therefore, the existence of well defined values for all the key classical properties (determinateness) cannot be denied. If determinateness prevails, however, then the mystery associated with QM dissolves under the realization that the theory is just incomplete. And so, following Einstein, the classical realist could claim that the reason why the motion of a proton must be described in terms of probabilities is only because some of the parameters which determine it are as yet hidden to us. Bohr, however, flatly rejected the notion that systems spread out in space were necessarily separable. Before the time of measurement -he maintained in Bohr (1935)- the experimental situation contemplated in the EPR argument cannot be understood in terms of separate entities but as an irreducible whole.

The result was a split in theoretical physics. A group of physicists favorable to classical realism initiated a series of ingenious attempts to *complete* QM by accounting for its probabilistic results in terms of various "hidden variables theories". The classical realists did not claim that QM was wrong; indeed, all the observable predictions of the theory they accepted as premises. The great majority of physicists, however, moved away from classical realism and proceeded under the assumption that QM was already a complete theory, giving birth to a physics famous for its unprecedented level of success in the history of science. The radical interpretation thus born, the Copenhagen Interpretation of the 1930's and its direct subsequent generalizations (relativistic QM, quantum electrodynamics, etc.), is rightly credited for having opened the door to a world of wonders whose scope and texture nobody had remotely anticipated.

Meanwhile, the hidden variables program continued to develop, but not without having to confront some telling difficulties, particularly von Neumann's famous proofs concerning the impossibility of the most straightforward version of hidden variables. Although these proofs forced theorists away from the early Einsteinian moulds, it was still possible to maintain that for every description in QM there was a "richer", more "intelligible" story in terms of hidden variables; yet, these theories were leading neither to better descriptions of the quantum phenomena, nor to new discoveries[2].

Hidden variables physics or Copenhagen QM? Success or intelligibility? There seemed to be no way to settle the dispute. Three decades elapsed without any agreement in sight; the question had become, in a sense, "metaphysical". Then, in 1964, Bell published his famous theorem.

3. BELL'S THEOREM

What Bell showed is that the EPR analysis does in fact lead to experimental predictions differing from those of QM, in particular that there is a limit on the extent to which certain distant events can be correlated, whereas QM predicts that under some circumstances this limit is exceeded.

The Bell theorem is both simple and subtle. Essentially, what it does is supplement the analysis of the original EPR argument by considering the instances of mixed "incompatible" measurements on the components of the total system. Applied to the simple case considered in this paper, the theorem proceeds basically as follows. The two protons in the singlet state are intercepted by two independent spin detectors, each containing an array of Stern-Gerlach magnets appropriate for measurements along three independent directions (i=1,2,3); a three-setting switch allows the experimenter to choose the orientation. For the sake of simplicity, we may think of the detectors as placed one to the right (R) and the other to the left (L). Since the spin components of individual protons are limited to two values (a=+,-), and following van Fraassen (1985), it is possible to study the situation in terms of four basic propositions:

1) Ri: Measurement done with detector R in setting i.
2) Lj: Measurement done with detector L in setting j.
3) Ria: Measurement done with detector R in setting i (1,2, or 3) with outcome a (+ or -).
4) Ljb: Measurement done with detector L in setting j (1,2 or 3) with outcome b (+ or -).

The EPR cases correspond, of course, to measurements done with the R and L detectors switched in the same setting. Using classical probability theory, the correlation and the principles specifically employed in the EPR argument may be described as follows:

I) EPR Correlation: The correlation can be derived either with the help of conservation principles, or taken as an experimental result; the claim is that the spins of the protons along any given direction i add up to zero:

$$P(Ria\&Lia \;/\; Ri\&Li) = 0$$

II) Principle of Locality: The world is made up of material systems governed exclusively by local interactions; in particular, all physical effects are propagated with subliminal velocities, so that no structure may be transmitted between regions separated by space-like intervals (no signals may be transmitted superluminary from one detector to the other); measurement results are therefore completely determined by the local enviroment of the measured system. Put in statistical terms:

$$P(Ria \;/\; Ri\&Lj) = P(Ria \;/\; Ri)$$
$$P(Ljb \;/\; Ri\&Lj) = P(Ljb \;/\; Lj)$$

III) Principle of Separability: This is a delicate issue concerning Bell's theorem, one about which much confusion exists. Whereas the Principle of Locality focusses on the impossibility of transmitting structure at superluminary speeds, the Principle of Separability focuses on the independence of the causal chains which lead to the measured values in the detectors. The notion of independent generation and development is thus central to the belief that the protons are able to hold their individuality, i.e., that their ontology is ultimately a separable one. If their respective chains are ultimately independent, one can think of the correlation between the protons as something causally traceable to the processess which take place in the preparation chamber (characterized here by the value q of a general variable A). In other words, the statistical correlations found in the laboratory are said to have a "common cause" (in Reichenbach's sense) in the past of the protons:

$$P(Ria\&Ljb/Ri\&Lj\&Aq) = P(Ria/Ri\&Lj\&Aq).P(Ljb/Ri\&Lj\&Aq)$$

Now, the thesis of Bell's theorem is that the above conditions entail certain inequalities, known as the "Bell inequalities". In particular, if we define $p(i,j) = P(Ri+\&Lj+ / Ri\&Lj)$, then the following expression is derived[3]:

$$p(1,2) + p(2,3) \geq p(1,3).$$

The important point is that QM entails the violation of this expression for certain particular arrangements, thus opening the way for an experimental resolution of the old debate initiated by Einstein and Bohr. The experiments inspired by this possibility are generically known as the "Bell experiments", and the science and technology used in them, especially over the last ten years or so, appears to be reliable beyond any reasonable doubt. The experimental findings so far are distinctly in favor of QM[4].

The conclusion appears to be, therefore, fairly unambiguous: the quantum mystery cannot be solved with the help of the classical notions of locality, separability and determinateness, for these notions lead to conclusions not borne out by experiment.

4. DIRECTING THE REFUTATIONAL ARROW

But then, once the technology of the Bell experiments is accepted as adequate, how are the results obtained with their help to be interpreted? The failure of the Bell inequalities shows that something philosophically fundamental is at issue. What that something is, however, depends on which assumption of the Bell theorem is finally blamed for the failure of the inequalities. It may be the Principle of Separability, or the Principle of Locality, or both; the Principle of Determinateness, being derivable from these two, is bound to be questioned one way or the other.

There are at least three major philosophical lines of response to the results of the Bell experiments: a) neo- positivism, b) scientific metaphysics, and c) scientific rationalism. A more general line would be logic, of course, but since the factors involved in the Bell experiments are both indefinitely numerous as well as exceedingly complex, it seems clear that what is primarily needed in this case is substantive -not formal- guidance.

a) Neo-positivism. The problematic character of quantum descriptions has led to a revaluation of naive observation among some influential thinkers. Neo-positivism refuses, or disowns, the notion that acceptance of a scientific theory involves the belief that it is true; acceptance -it is said- involves as belief only that the theory saves the phenomena. Accordingly, neo-positivits think of the challenges raised by QM as ultimately "ideological", and thus prefer to concentrate their analyses on the observable "surface" phenomena. Neo-positivism is an extremely problematic postion[5]. Yet, although philosophically excessive in many ways, it has been helpful in bringing to the attention of students of QM the generality of the Bell theorem.

b) Scientific Metaphysics. It is common for illustrious scientists to contribute with powerful speculations about the nature of physical reality and scientific understanding. Revisionary and often even surrealist, bounded only by the best background knowledge available, largely unconcerned with empirical success, these efforts might be classified under the name "scientific metaphysics". As such, scientific metaphysics is a highly speculative enterprise, conducted sometimes as part of a search for revolutionary theories, sometimes as a reaction to the alleged refutation of some particular principle deemed essential to "good" science. Virtually all the major protagonists of contemporary physics Einstein, Bohr, Bohm, Wheeler, Penrose, Hawking, Bell, to name but a few, have excelled also as scientific metaphysicians. Since the Bell experiments leave open several theoretical options, scientific metaphysics is nowadays particularly relevant to the discussion of QM. For the best background knowledge available leaves us with the choice of directing the refutational arrow either toward the Principle of Locality or toward the Principle of Separability. Trying to save the latter by accepting some form of action at a distance, for example, may go against the major corpus of scientific practice, yet this does not appear to be necessarily an incoherent move, and indeed some physicists are hoping to produce a coherent and "fully satisfactory" theory in this direction.

c) Scientific Rationalism. Finally, I want to describe the particular line followed in this paper, which is sometimes referred to as "scientific rationalism" because of its emphasis on the reasons for scientific change, as well as on the scientific revisability of claims at all levels (including the philosophical levels). It embodies a view of rationality at odds with both traditional rationalism and traditional empiricism, for it stresses the contingency and revisability of all beliefs and yet affirms the possibility of knowledge, basing this possibility on the contingent existence of beliefs which are both successful and free from *specific* doubts (as opposed to global, universal doubts). Science -the scientific rationalist claims- has taught us not only about the world but also how to learn and think about the world. The corpus of Shapere's recent general philosophy is particularly important in this direction, especially its explication of scientific acceptance and rejection in terms of rationality principles rooted in contingent substantive knowledge[6].

5. THE ACCEPTANCE OF QUANTUM ENTANGLEMENT

My propose in the rest of this paper is to rehearse a scientific rationalist response to the Bell experiments. How is the scientific rationalist to direct the refutational arrow after the Bell experiments? He may choose to avoid conflict with special relativity only by denying the separability of the protons before measurement. If that is his choice, no difficulties follow for the quantum theory, which -as Howard and others have pointed out- is, after all, a local, but non-separable theory; indeed, the notion of quantum non- separability (or "entanglement") helps us to understand why the quantum theory can explain the peculiar correlations observed in the Bell experiments[7]. But serious problems do ensue for those theories in which separability is a fundamental assumption, including general relativity. Or, the scientific rationalist may choose to avoid conflict with general relativity and endorse the Principle of Separability, a central element in our present understanding of the continuous point manifold on which the metric tensor (and so the basic ontology of the theory) is defined; in this case, the price would be the inclusion of something like action at a distance in the picture.

To the scientific rationalist the question is, of course, what reasons are there for choosing one way or the other. Neither the special nor the general theory suffer at present from any specific doubts other than those contributed by the Bell results themselves. The two theories differ, however, with respect to scientific success. Special relativity has proved vastly triumphant both within the quantum domain as well as beyond. Indeed, in the quantum domain the non-local structures of special relativity have been simply carried over to all quantum field theories and their generalizations, and so the special theory has been deployed in the microworld with flying colors. The success of general relativity, on the other hand, is more problematic: difficult to bring to bear on quantum phenomena, difficult to blend smoothly with quantum field theories, general relativity has led to much distinguished theorizing in quantum physics but not yet to scientific successes remotely like the ones found in relativistic quantum mechanics. Without disowning general relativity, therefore, it would seem fair to claim that presently the balance of success leans in favor of the special theory.

There is yet another important aspect, however. The scientific rationalist is also interested in the direction *specifically* selected by the actual debate which led to the Bell experiments (as opposed to the host of merely "possible" directions one might come to consider). The debate, we must recall, was generated by the clash between classical realism and the Copenhagen Interpretation. Here, however, the success of the Copenhagen Interpretation seems unquestionable, which directs the refutational arrow again toward the Principle of Separability. And so, both from the point of view of relativistic physics, as well as from the point of view of the specific debate which generated the Bell results, simple considerations of sheer success seem to point consistently against the notion of classical separability.

The scientific rationalist is therefore compelled to deny the universal separability and determinateness of physical systems. The protons in the singlet state -he comes to agree- are not separable entities, let alone entities endowed with spins! Nevertheless, his decision to drop the Principle of Separability can only be tentative; the post-Bell predicament is complex. On the one hand, the scientific rationalist feels compelled to move yet another step in the direction of the Copenhagen Interpretation, and in doing so confront a kind of radical physical holism at odds with the best classical intuitions. At the same time, however, he has clear independent indications that the Copenhagen Interpretation cannot be completely adequate, for it is a deeply paradoxical theory. In particular, he is reminded of Einstein's belief that the Principle of Separability provides the only imaginable objective criterion for the individuation of physical states, which -it intuitively appears- makes separability a necessary condition for the objective testing of physical theories. If, however, the accepted dynamics of the quantum state were compatible with the existence of *some* classical separability in the world, the problem about the Copenhagen Interpretation would largely disappear.The problem is that the dynamical law of quantum systems, which is so adequate for the explanation of the correlations studied in the Bell experiments, does not seem to allow for separability at *any* level.

The situation appears to be therefore as follows: in the light of the contextual considerations at his disposal, the scientific rationalist finds it appropriate to deny the Principle of Separability while accepting that his choice is problematic, contingent on present background knowledge, and not without certain duties. Conspicuously, one of these duties is to search for an adequate critical revision of general relativity[8].

6. MYSTERY RECOVERED

So far the scientific rationalist has managed to come to a conclusion with respect to what the Bell experiments primarily challenge (i.e., the universality of the classical concepts of separability and determinateness). What the experiments specifically confirm, on the other hand, is a lot less obvious. According to many, the Bell experiments simply put the Copenhagen philosophy beyond all reasonable doubt, an opinion delightfully voiced for instance by David Mermin:

The questions with which Einstein attacked the quantum theory do have answers; but they are not the answers Einstein expected them to have. We now know that the moon is demonstrably not there when nobody looks. Mermin (1981).

The Bell experiments clearly establish much less than that, of course, but Mermin does raise an important point. For these experiments would seem to bring back the mystery feared by Einstein, which indeed they do at least at two levels: first, in relation to the intelligibility of the quantum correlations; and second, far more deeply, in relation to the prospective universality of the Copenhagen Interpretation.

1) Some students of QM question the intelligibility of quantum entanglement. The Principle of Separability -they maintain- cannot be denied, for it is either an epistemological condition of adequacy, or a metaphysical principle, or a matter of definition. The problem with this objection is that it renders quantum correlations mysterious by a dubious contrast with the framework of classical metaphysics, without telling us why should the classical principles be granted special status. In Cordero (1988) a case is made concerning the scientific rise and fall of the "metaphysical" principles used in the EPR argument, with the conclusion that the classical notions of determinateness, separability and locality arise in fact from scientific findings slowly wrestled from nature, not from the discovery of any necessary conditions for all future discourse about nature. The classical principles -I there claim- were articulated one step at a time in the course of no less than three centuries, during which their scientific critique and rational generalization rarely stopped, to the point that the revision of their scope now prompted by the Bell experiments is nothing new in their history. Nor is there anything unprecedentedly mysterious about the notion of quantum entanglement per se: the world of quantum correlations is no more mysterious than the old Newtonian world of inert systems in gravitational interaction, or the Einsteinian world devoid of superluminary signals.

2) A second level of mystery is brought on by the prospect of the *pervasiveness* of quantum entanglement in nature. It is not known as yet how far from the truth is the Copenhagen Interpretation, of course, but the scientific rationalist must

agree that the Bell experiments move physics one step in that direction, which -as said before- unfortunately lands him in paradox. The problem with the Copenhagen Interpretation arises from the postulation of a single universal "linear" mode of evolution for the quantum state, still an unresolved issue after more than half a century of distinguished efforts. Why, then, we may ask, is such postulation maintained? There are several reasons for this, as we shall see shortly, but there is also a certain obsession with unity and simplicity, a matter worth of serious consideration; there is a distinct possibility that what might be at work here is an unwarranted metascientific urge to "resolve" the phenomena of quantum chance, measurement and all their probabilistic (or quasi-probabilistic) features in terms of a single universal dynamical law.

7. THE SEARCH FOR ALTERNATIVES

The problem with the Copenhagen version of QM as a universal theory of matter is that it faces the notorious "measurement problem", which is ultimately a paradox resulting from the clash between the universal simplicity of the dynamical evolution postulated by the Copenhagen Interpretation and what we actually observe around, with the result that, if this interpretation is correct, quantum physics and the enormous scientific and practical success of QM become a mystery.

Of course, it is always possible to defuse the mystery of measurement either by adopting an instrumentalist position, or by patching up the theory with an ad-hoc "projection postulate". Given that the second option is hardly of philosophical interest, are we then to conclude that the scientific rationalist is commited to an instrumentalist position with respect to QM? One thing seems certain: if he hopes for a realist interpretation, he is in trouble unless a reasonably interesting generalisation of the Copenhagen Interpretation can be found.

Traditionally, philosophers have turned to the formalistic program associated with von Neumann for help. Today, however, there is a growing feeling that this foundational program cannot really cope with the measurement paradox, despite the unusual levels of cleverness and hard work which have been (and still are) invested in it.

8. PHILOSOPHY AND WORKING-LEVEL PHYSICS

And so, in order to avoid paradox and save the possibility of realism, the scientific rationalist is compelled to reject the notion of a universal law of evolution for the whole of physics; yet, can he do this in a convincing way? One research area which suggests itself to him as a source of philosophical inspiration is, of course, ordinary scientific practice, "working-level physics" as opposed to the abstract and formalistic approaches of foundational studies. Working-level QM is atomic and nuclear physics, quantum chemistry, cosmology, etc. Even though it is the locus of practically all the scientific and technical success of QM, working level physics is a field largely disregarded by philosophers, perhaps not always without some reason[9]. Today, however, the philosophical relevance of working-level QM seems to be made increasingly clear by the combined impact of the Bell experiments and the failure of foundational studies to deal with the measurement paradox.

Does the realist have a chance then? I think a scientific rationalist may answer this question with a qualified "yes". There would be at least one condition: the realist must take the contemporary motivations for anti-realism very seriously, for the anti-realists do have a point. In particular, he must satisfy himself that the most interesting issues raised against realism do not provide compelling reasons against the view he is trying to defend. Only then can he forcefully charge that anti-realism simply fails to do justice to the unprecedented scientific and technical success of QM.

The approach proposed by the scientific rationalist focuses, therefore, on asking some old questions in a fresher way: What exactly is the theory that succeeds so dramatically in the quantum domain? Which descriptions, laws and explanations are truly taken seriously in working-level QM and why? Do physicists have compelling reasons for accepting them? In the opinion of the scientific rationalist, philosophers must try to approach these questions in terms far closer to science than has generally been the case (There are conspicuous exceptions, of course[10].

9. IS THERE A PHYSICS OF DISENTANGLEMENT?

Bearing in mind all the above points, let us turn to what has now become the central question for the scientific rationalist: Has physics anything philosophically interesting to say about how quantum superpositions come to an end?

A cursory examination of virtually any technical paper or textbook in applied quantum mechanics suggests that physicists and chemists have no shortage of prospective answers. Furthermore, these answers are not always ad-hoc or fictional. Typically, physicists appeal to laws and principles when they consider the termination of physical processes: the Heisenberg time-energy relation, the so-called "decay laws", the laws of quantum radiation, etc. The question is, therefore, not so much whether there is a quantum story about state disentanglement in the literature of physics, but whether the quantum narrative at the level of working-level physics can be made philosophically interesting through fair explication. How much of the story (or stories) about quantum disentanglement can be accepted, and why? To my knowledge no one has yet considered this issue in a comprehensive way. Yet seemingly articulate stories about quantum disentanglement appear at least at three major points in working-level QM, and so the scientific rationalist may have something like the following to get off the ground:

a) Energy transitions in both ordinary QM as well as quantum field theories. General features about these transitions are discussed in significant detail in the context of photon absorption/emission, as in, for example, Goldin (1982) or Loudon (1983). In a different (but related) way they are also discussed in the context of reaction mechanisms in both nuclear and particle physics, and quantum chemistry, as for instance in Shikorov (1982) and Pimentel (1969).

b) Proposals for a modification of the temporal law of QM. These comprise approaches such as 1) modifications of the law by physicists, as in Pearle (1983) or Ghirardi (1986); and 2) modifications motivated by chemical considerations, particularly approaches that grant irreversible processes a fundamental status over reversible processes, as in Prigogine (1980).

c) Studies of the temporal constraints imposed by the locality principle on "instantaneous" quantum transitions, a crucial matter, as in Fleming (1985), or -in a complementary way- Bunge (1985).

To the scientific rationalist, the interest of the above developments for the proper resolution of the mystery of quantum mechanics makes a philosophical explication of them imperative: What are the "laws of quantum transition"? What exactly do these laws amount to? What are the respective scopes and limits of the views summarized in (a), (b) and (c)? To what extent do these views suggest an integrated picture? Are quantum transitions a fundamental feature of the world, or are they derivative?

Another host of questions concern the testability of the type of theory envisaged here by the scientific rationalist. What if disentanglement proves impossible to test (in principle or in practice)? This philosophical issue is surely a relevant one. If the physics of disentanglement proves untestable but otherwise coherent, following the type of reasoning employed in his discussion of the Bell experiments, the scientific rationalist would be left with two physical theories to choose from (QM with and without objective disentanglement) which, although equally coherent from an internal point of view, would still radically differ with respect to their coherence with the rest of accepted knowledge.

If, on the other hand, the physics of disentanglement proves testable and fails, then quantum physics would move the scientific rationalist yet another step toward the Copenhagen Interpretation. If so, the persistence of the mysteries of superposition and entanglement might conceivably motivate him to try the relativist direction explored in Putnam (1983), or something along a radically instrumetalist or non-realist direction, perhaps as recommended in van Fraassen (1980). Surely the scientific rationalist might come to accept any of these options, but only if something like what is envisaged here happened. However, the culturally interesting point is that his options would then derive their warrant from the context of the highly internalized philosophy which the scientific rationalist has learned to articulate from the results of science and philosophy in the 20th century. Under some circumstances, therefore, science might indeed direct the scientific rationalist toward relativism or instrumentalism, but then the acceptance of either of these would be, as far as we can presently tell, a response *contingent both on what the scientific rationalist has learned about the world as well as on what he has learned about how to learn about the world*.

10. QUANTUM MECHANICS AND MYSTERY

However, if a realist resolution of the mysteries of QM were to prove possible after all, then -other things being equal- the case of general realism would be correspondingly strengthened within scientific rationalism. Needless to say, the realism envisaged here would be just as contingent on the ways of the the world as the anti-realism considered in the previous section. Yet the point is that, if the notion of quantum disentanglement were to prove acceptable, then there would be no specific reason for failing to take the quantum story seriously; in particular, the quantum model of reality (i.e., the picture of the world as something constituted, not primarily by individuals, but by quantum systems endowed with objective property-indeterminateness, entanglement and objective chance) would enjoy full coherence, utter success, and present freedom from specific doubts. What more can be reasonably demanded in order to accept QM?

Queens College, City University of New York.

NOTES

1 For a full presentation of the argument see Einstein et al. (1935), and Bell (1964).

2 For an excellent discussion of von Neumann's proof and its significance for hidden variables theories, see Bell (1966).

3 For the general derivation and commentary see, for example, van Fraassen (1982), Jarrett (1984), and Howard (1985).

4 For a discussion of the available technology and it reliability see Kamefuchi (1985); the most famous experiment to date is discussed in (Aspect 1982).

5 The leading defense of this form of empiricism is found in van Fraassen (1980); an apt application of this philosophy to the problems of QM is found in van Fraassen (1985). Some of the central problems with neo-positivism are discused, for example, in Suppe (1986) and Cordero (1987).

6 For a detailed presentation of these concepts see Shapere (1984), (1988).

7 For a perceptive discussion of this topic see Howard (1985).

8 I cannot possibly do justice to this topic here, but it seems clear that since no evidence of universal macroscopic quantum entanglement exists, the future theory must allow for something sufficiently close to the separable ontological structure of general relativity in the macroscopic limit. This simple requirement, however, might prove incompatible with the project of developing a universal physics. If so, the situation would confront the scientific rationalist with further - and perhaps deeper- choices. It is not impossible that he might try to save universality at all costs, should compelling reasons (by no means available at present) move him in that direction. Scientific metaphysicians are clearly needed in this connection if we are to be properly educated for the task when the time comes.

9 Working-level physics tends to look hopelessly unappealing to philosophers partly because it seems to have very little to add to what foundational studies already offer; moreover, it is full of rules of thumb, ad-hoc fictions, and so on, to the point that many leading "explanations" entertained in working-level physics have been rightly exposed in Cartwright (1983).

10 Indeed, although still a minoritarian approach, the work of Shapere and Shimony have been exemplary in this direction; moreover, in recent years some of the most distinguished younger philosophers and philosophical historians have began to take "working-level physics" very seriously, as in, for example, Cartwright (1983), Howard (1985), Wessels (1985) and Galison (1987).

BIBLIOGRAPHY

Aspect A. et al. (1982), "Experimental Test of Bell's Inequalities Using Time-varying Analizers". Phys. Rev. Letters 49: 1804

Bell, John S. (1964), "On the Einstein-Podolsky-Rosen Paradox". Physics 1: 195.

Bell, John S. (1966), "On the Problem of Hidden Variables in Quantum Mechanics". Reviews of Modern Physics 38: 447.

Bohr, Niels (1935), "Can Quantum Mechanical Description of Physical Reality Be Considered Complete?" Physics Review 48: 696.

Bunge M. and A.J. Kalnay (1983), "Real successive measurements on unstable quantum systems take nonvanishing time intervals". Nuovo Cimento 77B: 1.

Cartwright, Nancy (1983), How the Laws of Physics Lie. Oxford UP.

Cordero, Alberto:

(1987), "Observation in Constructive Empiricism". Critica, forthcoming.

(1988), "EPR and Compositionalism"; forthcoming.

Einstein, A., B. Podolsky and N. Rosen (1935), "Can Quantum- mechanical Description of Reality Be Considered Complete?" Physical Review 47: 777.

Fleming, G.N. (1985), "Towards a Lorentz Invariant Quantum Theory of Measurement". Workshop on Fundamental Physics, Univ. of Puerto Rico.

Galison, Peter (1987), How Experiments End. Chicago UP.

Ghirardi G.C. et al. (1986), "Unified dynamics for microscopic and macroscopic systems". Physical Review D 34:470.

Goldin, Edwin (1982), Waves and Photons. Wiley.

Howard, Don (1985), "Einstein on Locality and Separability". Studies in History and Philosophy of Science 16: 171.

Jarrett, J. (1984), "On the Physical Significance of the Locality Conditions in the Bell Arguments". Nous 18: 569.

Kamefuchi, S. et al. (1985), Foundations of Quantum Mechanics in the Light of New Technology. Physical Society of Japan.

Loudon, Rodney (1983), The Quantum Theory of Light. Oxford U.P.

Mermin, David (1981), "Quantum Mysteries for Anyone", Journal of Philosophy 78: 397.

Pearle, P. (1983), "Experimental Tests of Dynamical State-vector Reduction"; Physical Review D. 29: 235.

Pimentel G.C. et al. (1969), Chemical Bonding. Holden Day.

Prigogine, Ilya (1980), From Being to Becoming. Freeman.

Putnam, Hilary (1983), "Quantum Mechanics and the Observer"; Philosophical Papers, vol. III. Cambridge UP.

Shapere, Dudley (1984), "Objectivity, Rationality, and Scientific Change". PSA 1984 2: 637.

Shapere, Dudley (1988), The Concept of Observation in Science and Philosophy. Oxford UP, forthcoming.

Shimony, Abner (1987), "Search for a World View which can Accomodate our Knowledge of Microphysics"; presented at Notre Dame University, October 1987.

Suppe, Frederick (1986), "Nature and the Problem of Scientific Realism". Metaphysical Review. forthcoming.

Van Fraassen, B.C.:

(1980), The Scientific Image. Oxford UP.

(1985), "EPR: When is a Correlation not a Mystery?". Symposium on the Foundations of Modern Physics, edited by P. Lahti & P. Mittelstaedt. World Scientific.

Wessels, Linda (1985), "Locality, Factorability and the Bell Inequalities", Nous 19: 481.

Gino Tarozzi

PROBABILITY AND DETERMINISM IN QUANTUM THEORY

1. INTRODUCTION

At a variance with a popular opinion according to which the basic feature of modern physics would be reducible to its transition from classical determinism to quantum-mechanical indeterminism, I propose, rather, to put in evidence how the main peculiarity of quantum theory is constituted by an ambigous formal coexistence between a deterministic and a probabilistic description. In particular, I shall show that such an unsolved dualism can be considered an essential interpretative key to the main philosophical problems of this last theory.

In the first place it will be pointed out how the problem of measurement in microphysics has its logical roots in the presence of the two different kinds of evolution of the quantum mechanical state-vector: the former deterministic provided by the Schrödinger equation for unobserved physical systems, the latter probabilistic whenever a measuring operation takes place, without being given an unambiguos specification of the precise mutually exclusive conditions for these two types of transition. Then one of the most radical solutions corresponding to a complete reduction of quantum mechanics, to a strictly deterministic theory, via a causal completion of the description provided by the Schrödinger equation, is discussed, emphasizing the difficulties of this approach connected with the violation of the locality condition.

The dualistic nature of quantum formalism would seem to be the explanatory key also for the incompatibility between local real-

E. Agazzi (ed.), Probability in the Sciences, 237–259.

ism (or local hidden variable theories) and quantum mechanics. For, both the E.P.R. paradoxes and the Bell theorem after the refutation of its probabilistic proof, can be considered as consequences of the contrast within quantum formalism, between the possibility of making predictions with certainty about the values of two or more observables and the uncertainty principle according to which, since the previous observables are represented by non-commuting operators, they cannot have simultaneous reality. In such a perspective the residual determinism connected with the notion of predictability with certainty, which is anyhow a direct mathematical consequence of quantum mechanical description of correlated particles, might be regarded as a formal anomaly whose elimination in a properly probabilistic interpretation of quantum mechanics, would rule out the preceeding paradoxes. I demonstrate nevertheless that a probabilistic reformulation of the theory of spin $-\frac{1}{2}$ correlated systems is both logically and observably incompatible with a generalized reality principle in which deterministic predictability has been substituted by probabilistic predictability. Such a conclusion represents a confirmation of the thesis, already expressed, of the extreme seriousness of the basic problems of the foundations of modern physics which can be resolved neither from a merely philosophical, by replacing the Copenhagen with a realistic interpretation, nor from a strictly formal point of view, via the elimination of the ambiguous dualism of indeterministic vs. probabilistic description in quantum mechanics.

2. THE PROBLEM OF MEASUREMENT: DETERMINISTIC VS. PROBABILISTIC EVOLUTION OF THE STATE VECTOR

It has been emphasized by several authors and in particular by Wigner[1], that as far as the evolution of the state-vector, provided by the time-dependent Schrödinger equation is concerned, quantum mechanics appears as a strictly deterministic theory. In a certain way, it could be even said that the description of the state of a physical system given by the quantum-mechanical equation of motion is "more deterministic" than the one of classical mechanics. As a matter of fact in this last theory a small uncertainty in the definition of the initial conditions of a physical system, produces, after a sufficiently long time interval, an *arbitrarily large* difference in the final coordinates.

To this extent let us consider a classical macroscopic system constituted by a perfectly elastic ball moving along the direction x, between two rigid walls localted at $x = \pm l$. Let us moreover suppose that the initial position of the ball when thrown, can be defined only with an error Δx and similarly that an error Δp_x is made in the determination of the initial momentum p_x. This implies that, after a certain time t, we know the position of the ball with an error $\Delta x + \Delta p_x(t/m)$ which for $t \gg 0$ becomes even larger than the distance $2l$ between the two rigid walls. Therefore within the logical context of classical physics chance prevails as time elapses, to such an extent that, for an arbitrarily large time interval, the ball is completely devoid of space-localization.

On the contrary in quantum mechanics as far as one considers the equation of motion, if there is an initial uncertainty in the determination of the state-vector, this incertainty does not grow at any given successive time as a consequence of the unitary nature of the operator of time development. Nevertheless, unlike classical mechanics, quantum formalism is not reducible to an equation of motion describing the state of the system, but it supports such an equation with a theory of observables, entities which represent, together with the state of the system, the two basic notions of quantum theory. Now, while on one hand the equation of motion is strictly deterministic, i.e. once a state-vector $|\psi(t)\rangle$ is assigned at an arbitrary given instant $t = t_0$, then $|\psi(t \neq t_0)\rangle$ is completely and unambiguosly determined, on the other hand the theory of observables is closely probabilistic and allows only the prediction of the probabilities of finding certain results from measuring operations.

More precisely, given an observable $\mathscr{A}$ defined on a physical system $\mathscr{S}$ whose state $|\psi_i\rangle$, before the measurement, is described by the vector

$$(1) \qquad |\psi_i\rangle = \sum_{k=1}^{n} c_k |a_k\rangle$$

where $|a_k\rangle$ is the eigenvector of the linear hermitean operator A (associated to $\mathscr{A}$) relative to the eigenvalue a_k, the theory of observables allows the probability

$$|c_k|^2 = |\langle a_k|\psi\rangle|^2 \quad (2)$$

that $\mathscr{A}$ assumes the value a_k to be calculated. This means that if one finds the result a_k, as a consequence of the measuring process, the state of the system $\mathscr{S}$ has undergone a sudden modification consisting in the discontinue transition from the initial state of superposition $|\psi_i\rangle$ to a defined final state $|\psi_f\rangle = |a_k\rangle$.

Now, if we try to provide a mathematical treatment of the measuring interaction as for any other physical process, we have to attribute a state-vector also to the measuring device $\mathscr{M}$, so that the overall initial state of $\mathscr{S} + \mathscr{M}$ will be given by

$$|\psi_i'\rangle = \sum_{k=1}^{n} c_k|\alpha_0\rangle|a_k\rangle \quad (3)$$

where $|\alpha_0\rangle$ is the initial state-vector for $\mathscr{M}$. But as soon as we try to explain on the basis of the equation of motion the transition to the final state of $\mathscr{S} + \mathscr{M}$ and therefore apply, according to Schrödinger equation, the linear operator of time development U to the relation (3), we obtain for the final state

$$\begin{aligned} |\psi_f'\rangle &= U|\psi_i'\rangle = \sum_k c_k\, U\, \{|\alpha_0\rangle|a_k\rangle\} = \\ &= \sum_k c_k\, |\alpha_k\rangle|a_k\rangle \end{aligned} \quad (4)$$

which is a clearly unacceptable description since it contains a superposition of different macroscopic states for the measuring apparatus. To the extent of avoiding such a contradictory conclusion, it is necessary to postulate that, after every measurement, the apparatus records a well defined result. More generally given an ensemble of N physical systems whose initial state is described by $|\psi_i'\rangle$ one assumes that the measuring process, produces a transition to the proper mixture composed of

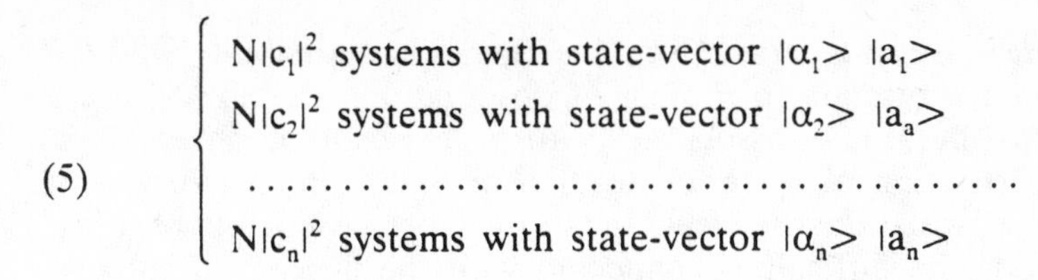

$$
(5)\quad \begin{cases} N|c_1|^2 \text{ systems with state-vector } |\alpha_1\rangle\, |a_1\rangle \\ N|c_2|^2 \text{ systems with state-vector } |\alpha_2\rangle\, |a_a\rangle \\ \dots\dots\dots\dots\dots\dots\dots\dots \\ N|c_n|^2 \text{ systems with state-vector } |\alpha_n\rangle\, |a_n\rangle \end{cases}
$$

where $|c_1|^2 + |c_2|^2 \ldots + |c_n|^2 = 1$ for the normalization condition.

The nature of quantum mechanical probabilism appears thus deeply differentiated from the one of classical mechanics. In fact whereas in this last theory chance prevails only when the change of the system with time is concerned, in the quantum one determinism continues prevailing also for the time development of unmeasured (or unobserved systems) to leave however space for probabilism as soon as the measuring processes take place. One is therefore faced with two different and incompatible modalities of evolution of the state-vector, the former deterministic for the ordinary physical processes, the latter probabilistic only for the measuring operations.

3. BOHM AND BUB'S DETERMINISTIC THEORY OF MEASUREMENT

To the extent of ruling out this ambiguos dualism from the logical structure of quantum theory, providing in this way a formal solution to the problem of measurement, one can try to reformulate quantum mechanics in such a way to riduce it to an entirely deterministic or, alternatively, to a properly probabilistic theory.

A complete probabilistic reformulation of quantum mechanics in the terms of a mere theory of observations has been proposed by Wigner. According to this author: "If we could not reduce the theory of observations to the theory of motion, we can try the opposite: we can eliminate the equation of motion and express all statements of quantum mechanics as correlations between observations"[2]. In Wigner's approach in the light of a radical phenomenistic positivism, the quantum-mechanical state-vector is completely deprived of empirical meaning, relegating it to the role of pure mathematical instrument useful for calculating the probabilities of the results of our observations or measurements (terms that in Wigner's use are perfectly equivalent). This possibility of attributing such a primate to the act of observation is realized once a projection operator is associated with each result of observation. Such a restriction of quantum formalism to the determination of statistical correlations between the results of subsequent observations avoids the difficulties connected with the process of collapse of the state-vector, mathematical entity practically banished in Wigner's approach.

Wigner's theory of observations has been critically discussed in detail elsewhere[3] and I shall limit myself here to mention the serious difficulty that it being consistent *only* as soon as the observations performed by *only one* observer are considered. In fact, if another person makes an observation the result of which he communicates to me afterwards, I am forced to treat him as a measuring apparatus and in this way to attribute to him a state-vector. But this means that after his observation and before I learn of the result found by him, his state will be paradoxically part of a linear combination of various observative results. Such a contradiction, known as the Wigner friend's paradox, can be eliminated only via the recourse to an interactionistic point of view with respect to the mind-body problem: to the extent of avoiding solipsism it is necessary to recognize the mind of the human beings and of course also the one of Wigner's friend, to be endowed with the faculty of generating the collapse of the state-vector. This last hypothesis, i.e. the assumption that the collapse represents a psycophysical event, seems however to be recently refuted by an experimental test[4]. In this way solipsism results the only solution logically consistent with Wigner's probabilistic theory of observations which appears therefore an adequate reformulation of quantum mechanics only in the light of a radical renouncement of the postulate of scientific objectivity and of explicit concessions towards an extreme version of conventionalistic relativism.

The diametrically opposite conceptual operation to reducing quantum formalism to a consistent deterministic theory, can be carried on in two different ways.

The most radical one consists in a complete elimination of the theory of observables, as the domain of quantum-mechanical probabilism, reducing quantum mechanics to its deterministic equation of motion. Such a possibility can be achieved by assuming that the very process of transition from the pure state (3) to the proper mixture (5) as a consequence of the measuring procedure, does never really take place and all the possibilities predicted by formalism would be therefore realized simultaneously in different branches of the universe.

This is the proposal contained in the theory of measurement of Everett[5], regarded by M. Jammer as "one of the most daring and most ambitious theories ever constructed in the history of science[6], in which the very conception of scientific theory as formalism *plus* interpretation by means of correspondence rules is entirely questioned by maintaining that "*the mathematical formalism of the quantum theory is capable of yielding its own interpretation*"[7].

The second, less ambitious attempt consists in the proposal to extend the deterministic structure of the equation of motion also to the behaviour of observables, and is explicitly contained in the theory of measurement of Bohm and Bub[8]. According to these authors such a purpose can be achieved if the description provided by the ordinary quantum-mechanical state-vector $|\psi>$ is completed via the introduction of hidden variables in such a way as to describe the measuring act as any other process of interaction between two physical objects: the measured system and the measuring apparatus.

With the aim of discussing briefly the theory of Bohm and Bub, let us consider, for semplicity, the case of a dicotomic variable $\mathcal{P}$ whose eigenstate are given by $|P_1>$ and $|P_2>$. The more general state-vector for the system S on which $\mathcal{P}$ is defined will be therefore:

$$(6) \qquad |\psi> = \psi_1|P_1> + \psi_2|P_2>$$

where the total angular momentum is

$$(7) \qquad J = J_1 + J_2 = 1$$

if one takes $J_1 = |\psi_1|^2$ and $J_2 = |\psi_2|^2$.

At this point in the theory of Bohm and Bub the description of S is assumed to be complete only if a second state-vector $<\xi|$ given by

$$<\xi| = \xi_1 <P_1| + \xi_2 <P_2| \tag{8}$$

is introduced.

What distinguishes $<\xi|$ from $|\psi>$ is the fact that while the former, like every ordinate state-vector of quantum mechanics has its components $\xi_1(t)$ and $\xi_2(t)$ obeying the deterministic Schrödinger equation, the new state-vector $<\xi|$ has complex components $\xi_1(t)$ and $\xi_2(t)$ characterized by a stochastic evolution. More precisely, in a four dimensional space the representative point of $\xi_1(t)$ and $\xi_2(t)$ is always on an ipersphere of unitary radius $\sum_i |\xi_i|^2 = 1$ with a density of probability uniformely distributed in the different points of this surface.In this way the overall evolution of the state of the system results deterministic as concerns the wave-function $|\psi>$ and stochastic as concerns the hidden variable $<\xi|$. More generally, given any quantum observable $\mathcal{R}$ its value is determined according to Bohm and Bub by some *non linear* function of the $|\psi>$, by the hidden variable $<\xi|$ and by the matrix R_{mn} associated to $\mathcal{R}$, which for an ensemble becomes

$$\overline{\mathcal{R}} = \int F(\psi, \xi, R_{mn})\ \varrho(\xi)\ d\xi \tag{9}$$

In the hidden variable theory of Bohm and Bub the measuring process is supposed to take place in the course of time intervals which can be considered small with respect to the characteristic time of variation of the hidden variables ξ_1 and ξ_2 ("impulsive" measurement) in such a way as the value of these parameters could be regarded as practically constant. Moreover one assumes that the equations of motion for ψ_1 and ψ_2 *during the measuring process* are:

$$\dot{\psi}_1 = \gamma(R_1 - R_2)\psi_1 J_2; \quad \dot{\psi}_2 = \gamma(R_2 - R_1)\psi_2 J_1; \tag{10}$$

where

$$R_1 = |\psi_1|^2 / |\xi_1|^2, \quad R_2 = |\psi_2|^2 / |\xi_2|^2 \tag{11}$$

and γ is a suitable positive real constant.

The relations (10) and (11) establish therefore a dependence of the time evolution of the two components ψ_1 and ψ_2 of the state-vector $|\psi>$ on the hidden variables ξ_1 and ξ_2. Moreover, if we multiply the first equations of (10) for ψ_1^* and the second one for ψ_2^*, where ψ_1^* and ψ_2^* denote respectively the complex conjugates of ψ_1 and ψ_2, we are led to

(12) $$dJ_1/dt = 2\gamma\ (R_1 - R_2)\ J_1J_2;\ dJ_2/dt = 2\gamma\ (R_2 - R_1)\ J_1J_2$$

and then summarizing it follows

(13) $$dJ_1/dt + dJ_2/dt = 0$$

This means that the equations (10) are consistent with a constant normalizations of $|\psi_1|^2 + |\psi_2|^2$ during the measuring process.

Let us then write the relations (12) as

(14) $$d(\log J_1)/dt = 2\gamma\ (R_1 - R_2)\ J_2;\ d(\log J_2)/dt = 2\gamma\ (R_2 - R_1)\ J_1$$

and consider the following two cases:
a) $R_1 > R_2$, with $J_2 \neq 0$: J_1 must always increase and J_2 must always decrease and because of the relation (7) this process will continue until $J_1 = 1$ and $J_2 = 0$, which implies in turn $\psi_2 = 0$. One can therefore conclude that in this case at the end of the measuring operation the ordinary state-vector $|\psi>$ evolves according to

(15a) $$|\psi> \rightarrow \exp i\varphi_1 |P_1>$$

where φ_1 is a phase factor.
b) $R_1 < R_2$, with $J_1 \neq 0$: (J_2 increases while J_1 decreases until $J_2 = 1$ and $J_1 = 0$. Then in a similar way the mathematical description after the measurement will be given by

(15b) $$|\psi> \rightarrow \exp i\varphi_2 |P_2>$$

where φ_2 is, like φ_1, a phase factor.

So it follows that the description given by the equations (10) produces the eigenstates $|P_1>$ or $|P_2>$ after a measurement of the observable has given the result + or − respectively. But, according to the theory of Bohm and Bub the result of a measurement of $\mathcal{P}$ is not only determined by the state-vector $|\psi\rangle$, which precludes the possibility of predicting such a result, but also by the hidden variable $\langle\xi|$ in a strictly causal way. In fact if $R_1 > R_2$, that is the ratio $|\psi_1|^2/|\xi_1|^2 > 1$ then the result will be +, while if $R_1 < R_2$, that is the ratio $|\psi_2|^2/|\xi_2|^2 > 1$, the result will be −. In this perspective, it is therefore only our present ignorance about the physical nature of the hidden variable $\langle\xi|$ which does not allow our deterministic prediction of the values of the observable $\mathcal{P}$.

2. NONLOCAL DETERMINISM

In Bohm and Bub's hidden variable theory a precise distinction between the ontological and epistemological level is achieved, assuming that physical reality evolves in a perfectly causal manner, independently from the knowledge of it possessed by the observer. According to such a perspective, endowed with considerable methodological fertility, the incognoscibility of the hidden variable $\langle\xi|$ is viewed as a contingent fact connected to the inadequacy of the experimental instruments employed in microphysics for its detection.

In the same way the wave-function $|\psi\rangle$ is regarded by Bohm and Bub "as playing a role somewhat analogous to the thermodynamic variables P, V, T" in classical statistical mechanics, corresponding respectively to the macroscopic variables pressure, volume and temperature. As a matter of course in this last theory, these large-scale thermodynamic properties are explained in terms of statistical distribution of atomic variables: "Relative to the large-scale level these variables are at first hidden, but they are *ultimately* observable with *new kinds* of experiments — Geiger counters or cloud chambers instead of thermometers and pressures gauges — which have in fact been developed *as a result of the new ideas* suggested by thinking in terms of atomic concepts"[9].

Such a methodological and epistemological fertility, common also to other hidden variable approaches, has been clearly pointed out by Evandro Agazzi, who has emphasized the *open realism* of these reinterpretations of quantum mechanics, which do not claim to exhaust the notion of reality within the limited sphere of scientific objectivity, i.e. within the domain of the predictable empirical properties of the physical object[10].

The reduction of quantum formalism to a deterministic description proposed in the theory of Bohm and Bub appears therefore as a conceptual operation satisfactory in a twofold way: both from the *formal* viewpoint since it eliminates the ambibuous coexistence between two radically different modalities of time-development of the state-vector, describing the measuring process through the very same mathematical law which regulates all other physical events, and from a more philosophical point of view, since it opposes an open realism to the orthodox interpretation of quantum mechanics, assuming that physical reality cannot be reduced to the mere observable properties which are experimentally detectable in a given theoretical context.

However, there exists a serious problem which appears as soon as such a theory is extended from the single system to two or more physical systems, and, as we shall see, which makes the theory of Bohm and Bub less appealing than ordinary quantum mechanics.

Regarding this, let us consider two spin $-\frac{1}{2}$ correlated systems P_1 and P_2whose quantum-mechanical state-vector is given by

(16) $$|\psi_s\rangle = \psi_1 \, |u_+\rangle \, |v_-\rangle - \psi_2 \, |u_-\rangle \, |v_+\rangle$$

where $|u_\pm\rangle$ and $|v_\pm\rangle$ are the state-vectors relative to the third spin component S_{13} and S_{23} of P_1 and P_2 respectively and where $\psi_1 = \psi_2 = 1/\sqrt{2}$ before any measuring operation takes place. In this case, the second state-vector $\langle\xi|$ can be expressed as

(17) $$|\xi\rangle = \left[\xi_1 \, |u_+\rangle + \xi_2 \, |u_-\rangle\right] \cdot \left[\eta_1 \, |v_+\rangle + \eta_2 \, |v_-\rangle\right]$$

where ξ_1 and ξ_2 are the hidden variables for P_1 while η_1 and η_2 are those for P_2.
Moreover, if such hidden variables are supposed constant during a measurement of spin, the relation (11), previously introduced, can be generalized as

(18) $$\dot{\psi}_1 = \gamma(R_1 - R_2) \, \psi_1 J_2; \; \dot{\psi}_2 = \gamma(R_2 - R_1) \, \psi_2 J_1;$$

where in a way perfectly analogous to the (11), $J_1 = |\psi_1|^2$ and $J_2 = |\psi_2|^2$, whereas R_1 and R_2 assume the different values

$$(19)\quad R_1 = 2(|\psi_1|^2 \,/\, |\xi_1|^2) + |\eta_1|^2;\ R_2 = 2(|\psi_2|^2 \,/\, |\xi_2|^2) + |\eta_2|^2;$$

Now, for $R_1 > R_2$ ($R_2 < R_1$) it follows from the (18) that ψ_1 (ψ_2) tends to 1 while ψ_2 (ψ_1) becomes 0, that is, we have the collapse of the state-vector, which after the measurement, will be given by

$$(20)\quad |\psi_f\rangle = \exp i\varphi_1 |u_+\rangle\, |v_-\rangle \quad (|\psi_f\rangle = \exp i\varphi_2 |u_-\rangle\, |v_+\rangle)$$

This mathematical treatment is valid for simultaneous measurements of the third spin component realized in space time-separated regions.

On the contrary if one performs *only* the measurement of the spin of P_1 (or P_2) the changes which are to be introduced in the preceedings description are not sensible. As a matter of fact it may be necessary to use a different constant γ in the (18) and to exclude the hidden variables η_1 and η_2 in the definition of R_1 and R_2. In any case one continues to obtain, as a result of the measurement, the deterministic collapse of the initial state of superposition given by $|\psi\rangle$ to a factorable state-vector like (20).

Nevertheless, such a collapse does not take place *only* for the state of P_1, but involves also the one of P_2, which assumes therefore *new properties* as a consequence of the measuring process realized on P_1. And this implies that the hidden variables ξ_1 and ξ_2, associated to P_1, together with the physical properties of the experimental device employed in the measurement performed on this system determine instantaneously at a distance a physical property of the system P_2. Therefore the explicitly nonlocal nature of the hidden variables introduced by Bohm and Bub in their theory of measurement emerges clearly. Their deterministic reduction of quantum formalism seems in this way to generate more problems than the ones it is able to solve, since the nonlocal consequences introduced by it in microphysics are automatically extended to macrophysics, because of the well known possibility of amplifying signals from the microscopic to the macroscopic domain, in open violation of the basic postulates of the relativity theory.

5. THE PROBLEM OF QUANTUM CORRELATIONS: PREDICTABILITY WITH CERTAINTY VS. THE UNCERTAINTY PRINCIPLE

We have seen in section 2. how quantum mechanics suffers a contradictory dualism due to an ambiguous coexistence between a deterministic equation of motion and a probabilistic theory of observables. This latter type of compresence, however, is not the only one within quantum formalism since there exists a second kind of dualism confined within the very domain of the theory of observables. Even this theory, as can be shown, does not appear to be entirely probabilistic in the description of some particular physical situations. As a matter of fact, let us consider two physical systems U and V which have interacted in the past and which, after the instant t_0, are spacetime-separated. The equation of motion permits us to calculate the wave-function $\psi(x_1,x_2)$ of the overall system U + V, where with x_1 and x_2 we have indicated respectively the position of U and V (to simplify our treatment we consider an unidimensional case). Let us then assume that U + V is found in a state described by the wave-function

$$(21) \qquad \psi(x_1,x_2) = \int \exp ip\,(x_1-x_2+x_0)\,dp$$

where x_0 denotes a constant. It can be shown that the wave-function (21) may be expressed in two different ways which are however equivalent from the mathematical point of view.

According to the first way, we can write $\psi(x_1,x_2)$ as

$$(22) \qquad \psi(x_1,x_2) = \int \psi_p(x_1)\,\varphi_p(x_2)\,dp$$

where

$$(23) \qquad \psi_p(x_1) = \exp ipx_1;\ \varphi_p(x_2) = \exp\left[-ip(x_2-x_0)\right]$$

are, respectively, an eigenstate for U relative to the momentum p and an eigenstate for V relative to the momentum $-p$.

According to the second way the (21) can be expressed instead as

$$(24) \qquad \psi(x_1,x_2) = \int u_x(x_1)\,v_x(x_2)\,dx$$

where

$$(25) \qquad u_x(x_1) = 2\pi\delta\ (x-x_1+x_0);\ v_x(x_2) = \delta(x_2-x_1)$$

are, respectively an eigenstate for U relative to the position $x-x_0$ and an eigenstate for V relative to the position x.

Suppose now a measurement of the momentum of U, which has given the result $p = \bar{p}$ is made. As a consequence of the collapse of the wave-function (22) one is led to the new wave-function

$$(26) \qquad \psi_{\bar{p}}(x_1)\ \varphi_{\bar{p}}(x_2)$$

One can therefore predict with certainty and without disturbing V in the assumption of the validity of the locality hypothesis, since at the time of the measurement on the physical system U, V is space-time-separated from U, that a measuring operation of its momentum will give $-\bar{p}$.

In a similar way if one performs a measurement of the position of U and finds $x_1 = \bar{x}_1$, that is $x = \bar{x}_1 - x_0$, from the collapse of the wave-function (24) one obtains

$$(27) \qquad u_{\bar{x}_1-x_0}(x_1)\ v_{\bar{x}_1-x_0}(x_2)$$

which allows one predict with certainty and without disturbance that the value of x_2 will be found equal to $\bar{x}_1-x_0$.

On the basis of this peculiar physical situation Einstein, Podolky and Rosen (EPR) proposed their famous paradox[11]. They showed, as is well known, that if one assumed that every physical theory has to satisfy a condition of completeness prescribing that all the empirical properties of the physical systems predictable with certainty and without disturbance — *real* according to their famous reality criterion — were to have a counterpart in the theoretical description, this involved that since quantum formalism allows the simultaneous prediction with certainty of the value of both the observables position and momentum defined on the unmeasured system V, the same formalism *then* must be able to describe them as characterized by such precise values.

Nevertheless this appears as a contradictory conclusion since quantum mechanism represents the previous observables by non-commuting operators and it can be easily shown that, in EPR's words, "when the operator (A and B) corresponding to two physical quantities ($\mathcal{A}$ and $\mathcal{B}$) do not commute they cannot have simultaneous reality"[16]. This means that the precise knowledge of $\mathcal{A}$ precludes an analogous knowledge of $\mathcal{B}$ and viceversa, i.e. it is neither possible to measure simultaneously $\mathcal{A}$ and $\mathcal{B}$, nor measure $\mathcal{A}$ and the $\mathcal{B}$ (or viceversa) with an *infinite* precision.

We know, from one of the axioms of quantum mechanics, that when an observable $\mathcal{A}$ has a precise value a, the physical system on which $\mathcal{A}$ is defined, is in a corresponding state $|a>$ which is an eigenvector of the operator A, associated with $\mathcal{A}$. If it were therefore possible to measure simultaneously $\mathcal{A}$ and $\mathcal{B}$ with infinite precision, obtaining as result a and b, because of the preceeding axiom, the state-vector should be after the simultaneous measurements, a common eigenvector of A and B relative to the eigenvalue a and b. And this is clearly absurd, since any two non-commuting operators do not allow common eigenvectors apart from exceptional cases, which do not include however the one of the operators representing the observables position and momentum.

Moreover it is useless even to measure first $\mathcal{A}$ and then $\mathcal{B}$. As a matter of fact, after the measurement of $\mathcal{A}$, the state of the system would be an eigenstate of A, which may be expressed as a superposition of the eigenstates of B. The second measurement would then produce the collapse of the superposition of several eigenstates of B to only one of them and since *every* eigenstate of B may be in turn expressed as a superposition of the eigenstates of A, the measurement of $\mathcal{B}$ leads to the loss of all other knowledge previously acquired about the value of $\mathcal{A}$.

Thus one obtain a paradoxical result according to which quantum formalism allows on one hand the simultaneous deterministic predictions of two empirical properties of a physical system and prevents on the other hand the attribution of these two real (according to the EPR reality criterion) properties to this physical system. The EPR logical contradiction is susceptible to being shifted to the empirical level without the additional assumption of the completeness of quantum mechanics[13] (we have seen anyway in section 3. how a causal completion of the quantum-mechanical description implies a violation of the locality condition).

In the following discussion we shall show that it is always the notion of the quantum deterministic predictability which stays at the basis of Bell's contradiction.

Let us consider two spin $-\frac{1}{2}$ correlated systems S_1 and S_2 in two different space-time regions R_1 and R_2 where the apparata M_1 and M_2 are located. M_1 and M_2 measure respectively the observables $\mathcal{A}$ and $\mathcal{C}$ finding the sequences of experimental results $\{A_i\}$ and $\{C_i\}$, all equal to $\pm$ 1.

At the point let us suppose to change suddenly the observable measured by M_2 and to measure $\mathcal{B}$ instead of $\mathcal{C}$, obtaining the new sequence of results $\{B_i\}$. In this way if M_1 would continue measuring $\mathcal{A}$, then for the local realist hypothesis, it would find the previous set of results

$$(28) \qquad \{A_i\} = r_{1i}\ (\mathcal{A}\mathcal{C}) = r_{1i}\ (\mathcal{A}\mathcal{B})$$

In fact since the change occurred in R_2 cannot have influenced the physical situation in R_1 one can attribute to the system S_1 the sequence $\{A_i\}$ as a real property of such a system. Of course the results of the measurement of $\mathcal{B}$, resorting to the previous notation, may be written as

$$(29) \qquad \{B_i\} = r_{2i}\ (\mathcal{A}\mathcal{B})$$

Let us now suppose, on the contrary, that the observable $\mathcal{B}$ had been measured on S_1 instead of $\mathcal{A}$, while M_2 measured $\mathcal{C}$. The the results $\{C_i\}$ would have been the *same* as in the experimental case, for the locality hypothesis already employed in writing the relation (28):

$$(28) \qquad \{C_i\} = r_{2i}\ (\mathcal{A}\mathcal{C}) = r_{2i}\ (\mathcal{B}\mathcal{C})$$

We ask now which succession of results would have been obtained in this case from the measurement of $\mathcal{B}$, that is, knowing r_{2i} $(\mathcal{A}\mathcal{B})$, to determine r_{1i} $(\mathcal{B}\mathcal{C})$. With the aim of supplying an answer to this question let us consider the quantum-mechanical description of the correlated system S_1 and S_2 whose initial state corresponds to $|\psi_s>$ given by relation (16).

Now if the two apparata are measuring the same observable $\mathcal{B}$ one can introduce the correlation function

$$(31) \qquad P(\mathcal{B}\mathcal{B}) = -1$$

This means that we can predict *with certainty*, given the result of the measurement of the observable $\mathcal{B}$ on the first system S_1, the result of the measurement of $\mathcal{B}$ on the second one.

In this way, since moreover the measurement on S_1 has not disturbed S_2 because of the locality condition, we can apply the reality criterion and write:

$$(32) \qquad r_{1i}(\mathcal{B}\mathcal{B}) = -r_{2i}(\mathcal{B}\mathcal{B})$$

Therefore for the relation (8) and (5) and the local realist hypothesis

$$(33) \qquad r_{1i}(\mathcal{B}\mathcal{C}) = r_{1i}(\mathcal{B}\mathcal{B}) = -r_{2i}(\mathcal{B}\mathcal{B}) = -r_{2i}(\mathcal{A}\mathcal{B}) = -\{\mathcal{B}_i\}$$

In this way one can write the tµ hree correlations functions

$$(34) \qquad \begin{cases} P(\mathcal{A}\mathcal{C}) = \frac{1}{n} \sum_{i=1}^{n} A_i C_i \\ P(\mathcal{A}\mathcal{B}) = \frac{1}{n} \sum_{i} A_i B_i \\ P(\mathcal{B}\mathcal{C}) = -\frac{1}{n} \sum_{i} B_i C_i \end{cases}$$

If we consider the relation

$$(35)\, P(\mathcal{A}\mathcal{B}) - P(\mathcal{A}\mathcal{C}) = \frac{1}{n} \sum_{i} A_iB_i - A_iC_i = \frac{1}{n} \sum_{i} A_iB_i(1 - B_iC_i)$$

we are led directly to the validity of the standard Bell inequality

$$(36) \qquad |P(\mathcal{A}\mathcal{B}) - P(\mathcal{A}\mathcal{C})| \leq \frac{1}{n} |A_iB_i(1 - B_iC_i)| = \\ = \frac{1}{n} \sum_{i} |A_iB_i| \cdot |1 - B_iC_i| = \frac{1}{n} \sum_{i} 1 - B_iC_i = 1 + P(\mathcal{B}\mathcal{C})$$

which, as is well known, can be violated by the predictions of quantum mechanics.

6. A PROBABILISTIC THEORY OF SPIN $-\frac{1}{2}$ CORRELATED PARTICLES

In the preceeding section we have shown that on the only grounds of the residual determinism contained in the notion of predictability with certainty, the EPR criterion of reality can be applied obtaining in this way both the Einstein and the Bell contradiction. With the aim of finding a proper way out for such a paradoxical situation, one could then try to eliminate any trace of determinism from the theory of observables transforming at least this part of quantal formalism in a properly probabilistic theory.

A similar purpose can be reached if one modifies the logical structure of quantum mechanics in such a way that none of our predictions on the properties of the unmeasured correlated system, realized on the grounds of the results found on the other one, is any longer deterministic and we could therefore only predict the probabilities of finding certain results from the second measuring operation.

In particular, concerning the peculiar case of the systems of spin $-\frac{1}{2}$ previously discussed, one will proceed by substituting the ordinary description given by the single state (16) with the one provided by the more general state-vector.

$$(37)\quad |\eta^{II}\rangle = \alpha\, |\psi_1\rangle\, |\varphi_1\rangle + \beta\, |\psi_1\rangle\, |\varphi_2\rangle + \gamma\, |\psi_2\rangle\, |\varphi_1\rangle + \delta\, |\psi_2\rangle\, |\varphi_2\rangle$$

where the terms ψ_1 and ψ_2 (φ_1 and φ_2) stay respectively for $|u_+\rangle$ and $|u_-\rangle$ ($|v_+\rangle$ and $|v_-\rangle$) contained in (16) and where α, β, γ,δ are nonnegative coefficients which satisfy the normalization condition.

Let us now suppose to have found $S_{13}=+1$ (in $\hbar/2$ units) from the measurement on the first system. As a consequence of the collapse of the state-vector, $|\eta^{II}\rangle$ will be transformed into the state $|\eta^{I}\rangle$ of the proper mixture

$$(38)\qquad |\eta^{I}_1\rangle = |\psi_1\rangle\, |\xi_1\rangle;\ |\eta^{I}_2\rangle = |\psi_2\rangle\, |\xi_2\rangle;$$

where the states $|\xi_1\rangle$ and $|\xi_2\rangle$ correspond respectively to the following linear combinations

$$(39) \quad |\xi_1 = 1 / \sqrt{|\alpha|^2+|\beta|^2}\,(\alpha\,|\varphi_1> + \beta \exp i\Phi\,|\varphi_2>)\,|\xi>_2$$
$$= 1 / \sqrt{|\gamma|^2+|\delta|^2}\,(\gamma\,|\varphi_1> + \delta \exp i\Phi'\,|\varphi_2>)$$

in which Φ and Φ' are arbitrary phases.

This implies that on the grounds of (38) we can *no longer predict with certainty* that the result of the second measurement of S_{23} will be -1 but *only the conditional probabilities of finding* $+1$ or -1, given $S_{13} = +1$, which will be defined respectively as

$$(40) \quad p(++) = |\alpha|^2/(|\alpha|^2+|\beta|^2); \quad p(+-) = |\beta|^2/(|\alpha|^2+|\beta|^2)$$

In a similar way, if the result of the first measurement had been -1 we would have been able to predict, as a consequence of the transition from $|\eta^{II}>$ to $|\eta^{I}_2>$ the probabilities

$$(41) \quad p(-+) = |\gamma|^2/(|\gamma|^2+|\delta|^2); \quad p(--) = |\delta|^2/(|\gamma|^2+|\delta|^2)$$

to find $S_{23}=+1$ and $S_{23}=-1$ respectively.

It is, of course, clear that in this new probabilistic context, where the notion of predictability with certainty has been banished, the EPR criterion of reality is *no longer applicable* and one does not arrive therefore at any kind of paradox. Such a possibility of eliminating the Einsten-Bell contradiction would moreover appear to represent a very general result, since it has been also shown that all the proof of Bell theorem for probabilistic local hidden variable theories contained an additional assumption not necessarily satisfied by these last theories[14]. And this implied that the only sure domain of validity of Bell theorem was restricted to deterministic theories of hidden variables or alternatively to deterministic local realism, as has been shown in the preceeding section.

One can ask therefore if by eliminating any residual determinism from the quantum theory of observables one could achieve a definite resolution of the problem of locality.

7. PROBABILISTIC REALITY AND THE GENERALIZED BELL THEOREM

It has been shown that *like quantum formalism also the reality criterion* – predictability as a sufficient condition of physical reality — can be generalized in a probabilistic fashion[14]. Such a generalization, which has been discussed in detail elsewhere, consists essentially in the following: if, at a time t, we can predict, no more with certainty, but only with a certain degree of inductive probability p, the value r of an observable $\mathcal{R}$, defined on a physical system S, instead of assuming that S has the property

$$(\pi) \qquad Vl\ (\mathcal{R}) = r$$

we shall limit ourselves to maintaining that S has the property

$$(\pi') \qquad P_r\ [Vl(\mathcal{R}) = r] = p$$

even before t ("Vl" and "Pr(Vl)") are respectively a functor and a functor of a functor denoting "the value of" and "the probability that the value of").

It has been moreover demonstrated how the introduction of this generalized principle, which contains the EPR reality assumption as a particular case — as a matter of fact when Pr $(Vl(\mathcal{R})=\underline{r})$ = 1 the property π' is reduced to π — involves that, given a statistical ensemble of spin $-\frac{1}{2}$ correlated systems their description must be provided by a statistical mixture of the states (38) also before the first or any other measurement takes place, in contrast with what is prescribed by quantum mechanics, according to which the previous physical systems must be described by the state vector (37) before any measurement is performed[15].

One is thus led to an EPR-type probabilistic contradiction, which can in turn, be transformed into an empirical contradiction by proving Bell theorem. To this extent let us introduce the correlation function relative to the measurement of the two observables A(a) and B(b) on P_1 and P_2, respectively, on the statistical mixture (37) which will be given by:

$$\begin{aligned} P(ab) &= \sum_i w_i \langle \eta_i^I | A(a) \otimes B(b) | \eta_i^I \rangle = \\ &= \sum_i w_i \langle A(a) \rangle_i \langle B(b) \rangle_i \end{aligned} \qquad (42)$$

where w_i is the probability for any state-vector $|\eta_i^I\rangle$ of the mixture (37), and where

$$(43) \quad \langle A(a)\rangle_i = \langle \psi_i | A(a) | \psi_i \rangle \; ; \; \langle B(b)\rangle_i = \langle \xi_i | B(b) | \xi_i \rangle$$

denote the expectation values of the observable A(a) and B(b) on the states $|\psi_i\rangle$ and $|\xi_i\rangle$ respectively.

It can be easily seen that the correlation function (42) leads to the validity of Bell's inequality without reporting the complete proof, which has been exposed elsewhere[15], if one observes its *deep formal analogy* with the correlation function for the theories of hidden variables.

$$(44) \qquad P(ab) = \int A(a\lambda)\, B(b\lambda)\, \varrho(\lambda)\, d\lambda$$

which represents the starting point of the ordinary proof of Bell's theorem. This analogy is so strict that the term w_i of relation (42) satisfies the same conditions of positivity and normalization of the density function $\varrho(\lambda)$ of (44).

On the contrary as has been demonstrated the correlation function for $|\eta^{II}\rangle$ results

$$(45) \qquad \begin{aligned} P(ab) &= \sum_i \sqrt{w_i}\, \langle \eta_i^I | A(a) \otimes B(b) | \eta_i^I \rangle = \\ &= \sum_i \sqrt{w_i}\, \langle A(a)\rangle_i\, \langle B(b)\rangle_i \end{aligned}$$

a relation devoid of formal analogy with the (44).

In particular it can be stressed that the term $\sqrt{w_i}$ is characteristic of the correlation function for a state vector of the second kind, which are *responsible* for the violations of Bell's inequality. It is therefore clear that even the Einstein-Bell contradiction cannot be solved by eliminating the causal anomaly generated by the concept of deterministic predictability within the general probabilistic theory of observables.

We can consequently conclude by stressing the essential role of this dualistic nature of quantum mechanics as an essential interpretative key, from the formal point of view, both with respect to the problem of measurement and to the one connected with the description of correlated systems. The acknowledgement of such a dualism means, at least, freeing oneself from those approaches to the foundations of quantum physics which have intransigently defended a presumed radical quantum probabilism, like in the case of the unconditioned acceptance of the results of the von Neumann theorem, which was regarded as proof of the mathematical impossibility of a causal completion of the intrinsic indeterministic quantum description of the physical world.

We have however shown that none of the formal modifications to the logical structure of quantum physics, previously discussed, is able to provide an adequate solution to these basic logical and epistemological problems. This sheds a new light on the particular gravity of quantum paradoxes, originating from the fact that they not only arise in relation to the very structure of quantum formalism, and not merely as regards to its interpretation, but they cannot even be eliminated by means of integrations, or partial *ad hoc* changes of formalism itself.

Institut of Philosophy, University of Urbino

REFERENCES

[1] Wigner, E.P. (1971), "The Subject of Our Discussion", in *Foundations of Quantum Mechanics*, Rendiconti SIF, IL, Academic Press, New York.

[2] Wigner, E.P. (1973), "Epistemological Perspectives on Quantum Theory", in *Contemporary Research in the Foundations and Philosophy of Quantum Theory*, ed. by C.A. Hooker, Reidel, Dordrecht, p. 375.

[3] Tarozzi, G. (1981), "The Theory of Observations, Wigner Paradox and the Mind-Body Problem", *Epistemologia*, IV, p. 37.

[4] Hall, J., Kim, C., McElroy, B. and Shimony, A. (1977), "Wave-Packet Reduction as a Medium of Communication", *Foundations of Physics*, 7, p. 759.

[5] Everett III, H. (1957), "'Relative States' Formulations of Quantum Mechanics", *Reviews of Modern Physics*, 29, p. 154.

[6] Jammer, M. (1974), *The Philosophy of Quantum Mechanics*, Wiley, New York, p. 517.

[7] De Witt, B.S. (1971), "The Many-Universes Interpretation of Quantum Mechanics", in *Foundations of Quantum Mechanics*, Rendiconti SIF, IL, Academic Press, New

York, p. 212. A critical discusion of Everett's approach is contained also in G. Tarozzi (1985), "Teoria e strumento in microfisica", *Epistemologia*, VIII, p. 83.

[8] Bohm, D. and Bub, J. (1966), "A Proposed Solution of the Measurement Problem in Quantum Mechanics by a Hidden Variable Theory", *Reviews of Modern Physics*, 38, p. 453.

[9] Bohm, D. and Bub, J. *ibidem*, p. 457.

[10] Agazzi, E. (1969), *Filosofia della Fisica*, Manfredi, Milano.

[11] Einstein, A., Podolsky B., and Rosen, N. (1935), "Can Quantum-mechanical Description of Physical Reality Be Considered Complete?", *Physical Review*, 47, p. 777.

[12] Einstein, A., Podolsky B., and Rosen, N., *Ibidem*, p. 778.

[13] Tarozzi, G. (1981), "Local Realism and Bell's Theorem Without the Hidden Variables Hypothesis", *Atti dell'Accademia delle Scienze di Torino*, 108, p. 119.

[14] Selleri, F. and Tarozzi, G. (1980), "Is Clauser and Horne's Factorability a Necessary Requirement for a Probabilistic Local Theory?", *Lettere al Nuovo Cimento*, 29, p. 533.

[15] Selleri, F. and Tarozzi, G. (1983), "A Probabilistic Generalization of the Concept of Physical Reality", *Speculations in Science and Technology*, 6, p. 55.

INDEX OF NAMES

INDEX OF SUBJECTS